AF443276

POSTHARVEST GRAIN LOSS ASSESSMENT METHODS

A Manual of Methods for the Evaluation of Postharvest Losses

developed and compiled by

Kenton L. Harris
and
Carl J. Lindblad

ISBN 0-913250-14-7

1978

published in cooperation with
The League for International Food Education
The Tropical Products Institute (England)
Food and Agriculture Organization of the United Nations
Group for Assistance on Systems Relating to Grain After-Harvest

by the
American Association of Cereal Chemists

under Grant AIB/ta-G-1314
Office of Nutrition, U.S. Agency for International Development

Cover figure: British Crown Copyright. Reproduced with permission of the Controller of Her Britannic Majesty's Stationery Office.

CONTENTS

This volume stems from the joint and
independent efforts of many who have
contributed ideas as well as manuscripts.

When world food production is viewed as a system, loss and deterioration is
seen as a major food-limiting factor. Postharvest loss reduction would benefit
from reliable loss estimates and cost/benefit comparisons; improvements
also must be acceptable and feasible to introduce.

Determination of postharvest grain losses requires a blending of, and concepts from, several sciences.

         Postharvest, losses, food, insects, microbiological defined.

         Project planning involves many disciplines and concepts, from national
         priorities to logistics and local cultural values.

         Determination of losses should proceed stepwise from understanding
         the overall grain-food pipeline to location of leaks and sites where
         losses are relatively important, can be assessed, and are amenable to
         loss-reducing interventions.

         Rapid on-site appraisals (30–60 days) are both workable and useful to
         determine feasibility for further investigations and for some inputs, and
         to delineate specific problem points.

         Grain loss does not exist independent of human and social influence.
         Loss assessment and reduction programs need to be seen from within
         the local setting. Cross-cultural sensitivity and understanding are
         essential in planning and executing such efforts. Reminders are given
         on who, what, and how to obtain reliable, useful information on and
         within the social and domestic organizations and in relation to
         individuals. Special attention is given to the role of women.

Basic statistical requirements for surveys, sampling, probabilities, and other
concepts required in the assessment of losses are presented.

CONTRIBUTORS AND AUTHORS

The scope and format of this manual stem from the Technical Advisory Committee of the American Association of Cereal Chemists and from two meetings. One was held Sept. 8-10, 1976, at Harpers Ferry, WV. The other was held June 13-24, 1977, at the Tropical Stored Products Centre, Slough, England. Those present at these meetings are as much contributors as are those who eventually wrote the individual sections.

The 1976 meeting was a wide-ranging brainstorming session covering the basic concept of the manual and getting to the fundamentals of feasibility, format, and scope. It was a group effort and the benefits stemming from its interdisciplinary makeup cannot be overemphasized.

The 1977 meeting was a technical workshop devoted to defining and clarifying general goals and specific subjects and writing them down. It functioned both as a group effort and as a vehicle for individual contributions.

The American Association of Cereal Chemists Committee consisted of Edith A. Christensen. U.S. Department of Agriculture, Inspection Division, FGIS, Washington, DC 20250; John H. Nelson. (now) American Maize Products Company, Hammond, IN 46336; and Raymond J. Tarleton. American Association of Cereal Chemists, 3340 Pilot Knob Road, St. Paul, MN 55121.

The consulting-editing relationship with Hugh J. Roberts of L.I.F.E. and with Peter Tyler, Tropical Stored Products Centre, warrant special mention.

Credit is given to the El Salvador Centro Nacional de Technologie Agropecuaria (CENTA) for providing field and laboratory assistance in evaluating portions of this manual.

Participants at the two meetings and authors are given in the lists that follow.

PARTICIPANTS AT THE POSTHARVEST GRAIN LOSSES METHODS WORKSHOP

Harpers Ferry, WV
September 8-10, 1976

J. Mervyn Adams. (now) The Wellcome Foundation, Ravens Lane, Berkhamsted, Herts., England

Keith Byergo. Crop Protection, Office of Agriculture, Bureau of Technical Assistance, Agency for International Development, Washington, DC 20523

Howard R. Cottam. Consultant, 2245 46th St. N.W., Washington DC 20007

M. G. C. McDonald Dow. Board of Science and Technology for International Development, National Academy of Sciences, 2101 Constitution Ave., Washington, DC 20418

Maryanne Dulansey. Consultants in Development, 298 West 11th St., New York, NY 10014

Kenton L. Harris. Consultant, 7504 Marbury Road, Bethesda, MD 20034

William J. Hoover. American Institute of Baking, Box 1448, Manhattan, KS 66502

Henry Kaufmann. Cargill, Inc., Box 9300, Minneapolis, MN 55440

Carl Lindblad. Consultant, 1706 Euclid St. N.W., Washington, DC 20009

Floyd E. O'Quinn. 7328 Range Road, Alexandria, VA 22306

Priscilla Reining. International Office, American Association for the Advancement of Science, 1515 Massachusetts Ave. N.W., Washington, DC 20005

Hugh J. Roberts. League for International Food Education, 1126 16th St. N.W., Washington, DC 20036

PARTICIPANTS AT THE SLOUGH, ENGLAND WORKSHOP ON POSTHARVEST GRAIN LOSS METHODOLOGY

June 13-24, 1977

J. Mervyn Adams. (now) The Wellcome Foundation, Ravens Lane, Berkhamsted, Herts., England

Bill Andrews. TPI (TSPC), London Road, Slough, Berks, England SL3 7HL

Andy Baker. TPI (TSPC), London Road, Slough, Berks, England SL3 7HL

Robin Boxall. Institute of Development Studies, University of Sussex, Brighton, Sussex, England

Geoffrey G. Corbett. FAO, Via delle Terme di Caracalla, 00100 Rome, Italy

David Dendy. TPI, Industrial Development Department, Culham, Abingdon, Oxon, England

Jacques Deuse. IRAT, B.P. 5035, Montpellier, France

Bruce Drew. Pillsbury Company, 311 2nd St. S.E., Minneapolis, MN 55414

David Drummond. Ministry of Agriculture, Fisheries and Food, Pest Infestation Control Laboratory, Tolworth, Surrey, England

Rennie Friendship. TPI (TSPC), London Road, Slough, Berks, England SL3 7HL

Peter Golob. TPI (TSPC), London Road, Slough, Berks, England SL3 7HL

Martin Greeley. Institute of Development Studies, University of Sussex, Brighton, Sussex, England

Geoffrey Harman. TPI, 56/62 Gray's Inn Road, London WCIX 81U, England

Kenton L. Harris. AACC/L.I.F.E., 7504 Marbury Road, Bethesda, MD 20034

Noel Jones. TPI, 56/62 Gray's Inn Road, London WCIX 81U, England

Carl Lindblad. AACC/L.I.F.E., 1706 Euclid St. N.W., Washington, DC 20009

Matthias Von Oppen. ICRISAT, Hyderabad, India

Elizabeth Orr. TPI, 56/62 Gray's Inn Road, London WCIX 81U, England

Harry Pfost. Department of Grain Science and Industries, Kansas State University, Manhattan, KS 66506

Peter F. Prevett. TPI (TSPC), London Road, Slough, Berks, England SL3 7HL

Barbara Purvis. ESHH, FAO, Via delle Terme di Caracalla, 00100 Rome, Italy

Eberhard Reusse. FAO, Via delle Terme di Caracalla, 00100 Rome, Italy

Robert A. Saul. 1412 Martin Road, Albert Lea, MN 56007

Gerard G. M. Schulten. Royal Tropical Institute, 63 Mauritskade, Amsterdam-Oost, Netherlands

Harlan Shuyler. FAO, Via delle Terme di Caracalla, 00100 Rome, Italy

Philip Spensley. TPI, 56/62 Gray's Inn Road, London WCIX 81U, England

Malcolm Thain. TPI, 56/62 Gray's Inn Road, London WCIX 81U, England

Peter Tyler. TPI (TSPC), London Road, Slough, Berks, England SL3 7HL

David Webley. TPI (TSPC), London Road, Slough, Berks, England SL3 7HL

AUTHORS

J. Mervyn Adams. (now) The Wellcome Foundation, Ravens Lane, Berkhamsted, Herts., England

Geoffrey G. Corbett. FAO, Via delle Terme di Caracalla, 00100 Rome, Italy

David Dendy. TPI, Industrial Development Department, Culham, Abingdon, Oxon, England

Bruce A. Drew. The Pillsbury Company, 311 2nd St. S.E., Minneapolis, MN 55414

P. Golob. TPI (TSPC), London Road, Slough, Berks, England, SL3 7HL

Theodore A. Granovsky. Department of Entomology, Texas A & M University, College Station, TX 77843

John H. Greaves. Pest Infestation Control Laboratory, Tolworth Surbiton, Surrey, England

Martin Greeley. Institute of Development Studies, University of Sussex, Brighton, Sussex, England

Allan Griff. 5324 Wakefield Road, Bethesda, MD 20016

Geoffrey W. Harman. TPI, 56/62 Gray's Inn Road, London WCIX 81U, England

Kenton L. Harris. 7504 Marbury Road, Bethesda, MD 20034

William J. Hoover. American Institute of Baking, Box 1448, Manhattan, KS 66502

William B. Jackson. Bowling Green State University, Bowling Green, OH 43403

Henry Kaufmann. Cargill, Inc., Box 9300, Minneapolis, MN 55440

Carl J. Lindblad. 1706 Euclid St. N.W., Washington, DC 20009

Guy Martin. I.T.C.F. Cereal Laboratory, 46 rue de la Cleff, 75005 Paris, France

Jean-louis Multon, Institut National de la Recherche Agronomique, 44072, Nantes Cedex, France

T. N. Okwelogu. Produce Inspection Headquarters, PMB 1012, Enugu, Anambra State, Nigeria

Conrad C. Reining. Department of Anthropology, The Catholic University, 620 Michigan Ave. N.E., Washington, DC 20011

Eberhard Reusse. FAO, Via delle Terme di Caracalla, 00100 Rome, Italy

Robert A. Saul. 1412 Martin Road, Albert Lea, MN 56007

Gerard G. M. Schulten. Royal Tropical Institute, 63 Mauritskade, Amsterdam-Oost, Netherlands

Manfred Temme. Environmental Studies Center, Bowling Green State University, Bowling Green, OH 43403

POSTHARVEST GRAIN LOSS ASSESSMENT METHODS

PREFACE

When world food is viewed in terms of a *system* of production, distribution, and utilization, it becomes obvious that in our attempts to improve the system we have allocated most of our resources to the production component. Distribution and utilization have been comparatively neglected. But hunger and malnutrition can exist in spite of adequate food production. They can be the result of unequal distribution of food among nations, within nations, within communities, and even within families. Loss and deterioration of available food resources further adds to the problem. Hence, maximum utilization of available food is absolutely essential.

Of the agricultural commodities consumed as food, grains (cereals, legumes, oilseeds) contribute the bulk of the world's calories and protein. The food grains system is depicted in Fig. 1, which shows the many points at which losses of food occur. The reduction of postharvest grain losses, especially those caused by insects, microorganisms, rodents, and birds, can increase available food supplies, particularly in less developed countries where the losses may be largest and the need is greatest.

In September 1975, the growing international awareness of the need for reducing postharvest food losses culminated in a resolution of the Seventh Special Session of the United Nations General Assembly stating that "the further reduction of post-harvest food losses in developing countries should be undertaken as a matter of priority with a view to reaching at least 50% reduction by 1985." Yet, following the Seventh Special Session, an Interdepartmental Subcommittee reviewed past and current activity and concluded: "There is no agreed methodology of post-harvest loss assessment. Moreover, loss data are generally unrelated to the cost of loss reduction."

In its interpretation of available information on losses, the Subcommittee concluded that "there can be no agreed single figure for the percentage of post-harvest losses on a global scale or even on a national basis. There is clearly a need for more accurate assessment of these losses, to establish firm justification for the development and introduction of measures designed to reduce them where the cost/benefit ratios of corrective measures are favorable."

The goal of this volume is to provide postharvest grain loss assessment methods yielding standardized and reproducible results so that effective grain loss reduction efforts can be undertaken in developing countries. The assessment information from such a manual may provide essential justification and motivation for introducing measures designed to reduce grain losses.

This volume is prepared in large part for use by policymakers who need loss information both in determining national priorities and requirements and in bringing their efforts to bear on the small farmer and other small-volume grain handlers. It is also directed to the individual investigator who seeks a basic guide in his specific investigations. The manual is aimed primarily at loss assessment in developing countries.

Although a methodology for assessing postharvest grain losses will not in and of itself reduce those losses, the methodology is essential to postharvest operational programs so that priorities for loss reduction can be determined. In addition to serving as a much-needed assessment tool, the methodology and other activities proposed can serve as a means to persuade all concerned that change is necessary and that effective techniques for reducing losses are available. Even financial constraints can disappear when priorities are reordered.

As detailed later in this Preface and in Chapter II, the enormous variability of local postharvest situations dictates that no complete or definitive loss assessment methodology for all situations is now possible. Thus, this edition is not proposed as a final and absolute piece of work. For example, there exists very little experience which can be drawn from in loss assessment of cereal grains such as sorghum, millet, teff, and major oilseeds. Judgment will be

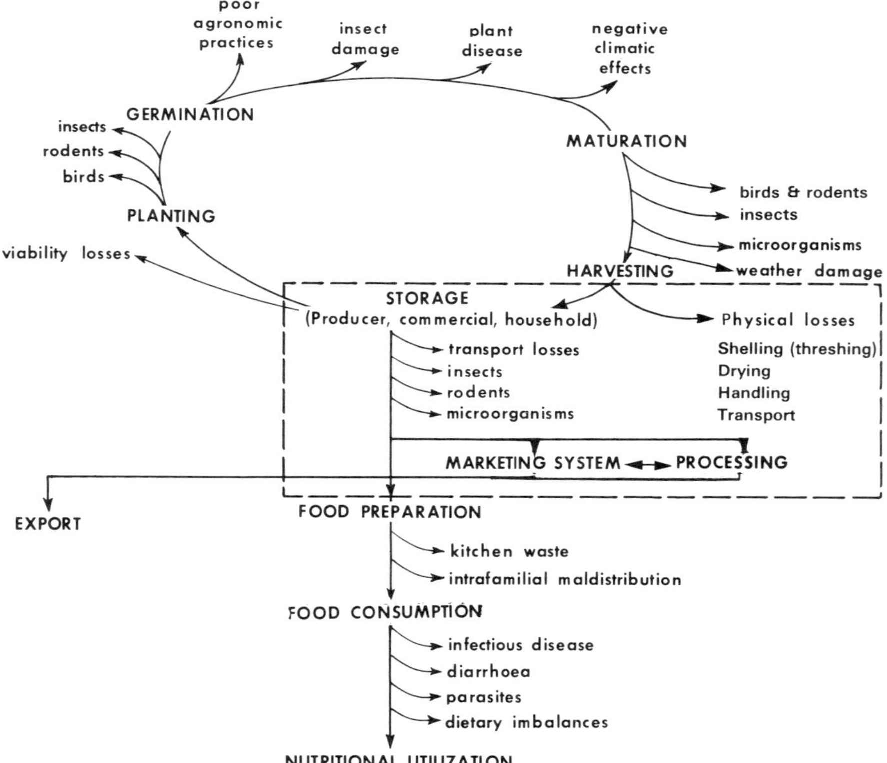

Fig. 1. Losses in the food grains system (farm level only). Box indicates focus of this manual.

required to adapt known assessment methods to those grains and to other loss situations not dealt with in sufficient depth here. Further, the editors realize that expansion and refinement of the loss assessment techniques presented in this manual are desirable and necessary as a continuing process.

Increasing food production by increasing acreage or yield per acre has been a readily applied concept while reducing losses to increase food supplies was a less obvious strategy. This occurred in spite of the availability of a considerable body of information on postharvest grain losses, and in spite of several decades of research and development on losses and their control.

Progress in reducing postharvest food losses requires the identification and elimination of the constraints to the application of existing technology. The major constraint may be a lack of finances, but it is equally possible that lack of knowledge and of trained personnel, as well as political and cultural constraints, exists. In 1975 an FAO Subcommittee position paper identified four constraints to the effective use of available technology for reducing on-farm losses: 1) lack of arrangements for producing the necessary inputs, 2) inadequate distribution channels for the necessary inputs, 3) lack of purchasing power or credit arrangements for the farmer to buy the inputs, and 4) inadequate information to the farmer on how to use the inputs.

While calling for integrated country programs to address these constraints, the Subcommittee stressed the need for creating "an awareness throughout national extension services that on-farm losses are serious and can be significantly reduced." Postharvest loss reduction intervention must be made, however, with specific techniques applied to reduce specific losses. While there may be broad sweeping national needs, not only are the techniques specific, but they must be applied at specific intervention points. Until data are available to show the potential gain from the elimination of losses amenable to reduction, motivation to reduce those losses will not be strong. But aggregate data reflecting losses on a global or even on a national basis are not really useful even if it were possible to obtain them. They are singularly unpersuasive to a farmer, trader, or warehouseman who must lay out his money and time.

Losses vary by crop, variety, year, pest and pest combination, length of storage, methods of threshing, drying, handling, storage, processing, transportation and distribution, rate of consumption, and according to both the climate and the culture in which the food is produced and consumed. Given such enormous variability, it is not surprising that reliable statistics regarding the type, location, causes, and magnitude of postharvest grain losses are not available. Yet reliable and objective methods for generating them are needed if priorities are to be given to the reduction of losses. This is needed in regional and national planning and in motivating those organizations which may fund loss-reduction programs, and on down to the local level.

Meanwhile, it is prohibitively expensive and unjustifiable to mount country-wide assessment studies of losses in the whole postharvest system. As detailed in Chapter II, an expert judgment is needed to identify the most serious grain loss points in a country's postharvest food supply system in order to mount in-depth assessment efforts at those high loss points.

Stated another way, changes will not be widely accepted until and unless they are practical for and clearly benefit the individual who is to make the change. Although losses and savings are far from the only elements which must be considered in loss reduction efforts, reliable figures can go a long way in convincing those dealing with grain, and certainly for motivating those organizations which may fund the loss reduction programs.

Extent of loss is important, but not all-important. Other factors should be considered in deciding on the nature of interventions, or whether to intervene at all: The value of the grain in economic lines; the fact that there will be social change effected by intervention programs; competition or conflict, or both, with other national priorities; effect on price stability and similar economic considerations; the relationship and possible conflict of economic factors that affect the consumer, grain grower, grain trader, and national balance of payments mean that interventions need to be subjected to an integrated, multidisciplinary evaluation and actually field tested within the social and economic structure before they should be implemented on a broad scale.

Both "guesstimates"[1] by knowledgeable people and estimates without factual basis, particularly by people with vested interests, have had a useful role in the past, will continue to be used in the future, and are especially useful when timely opinions are needed as to where the more serious losses occur. In using guesstimates to justify cost/benefit comparisons or to reshape established practices, however, one needs to recognize the possible bias of the estimator: Was it put in perspective by a thorough gleaning of the information, was the judgment based on an in-depth and long-standing knowledge of local or even country-wide conditions, was it made to reveal some situations and cover others? It is critical to understand that guesstimates are the type of estimations that requires the most *expert* judgment.

If large area or national survey figures are taken without sufficient regard for variations in the individual components, these figures may not be useful to locate specific intervention points.

Finally, we might ask why, in the face of a need for accurate figures that has not gone unnoticed over at least two decades, have there been so many postharvest loss estimates made with obvious biases, and why has a methodology not been forthcoming from the scientific community?

As stated above, the guesstimates have served a useful purpose. They have also been accepted by those seeking national resources and changes as well as by those allocating international resources. Although the scientific need was there, the political- and transformation-related requirements did not call for scientifically derived figures. Now, with increased sophistication and increasingly limited resources requiring benefit-related priorities, there is a need to know what the postharvest losses really are. Without such information, it is impossible to assess needs or to calculate improvements. However, there has been another factor that has stood in the way of assembling this manual. It needs to be mentioned, for its recognition is the key to the present status and ultimate fate of this volume. This factor has been the simple absence of anyone to do the job.

Within the L.I.F.E. consortium, the American Association of Cereal Chemists, under a contract from the Office of Nutrition, Technical Assistance Bureau, U.S. Agency for International Development, has broken the impasse on

---

[1]This term is used to connote estimates with some facts by knowledgeable people.

how and by whom the job was to be done, and it has developed and printed this volume with the hope that it is a volume to be evaluated, tested, and improved by actual use in the field. We look forward to the inevitable changes.

Kenton L. Harris
Carl J. Lindblad
August 1978

I. INTRODUCTION

K. L. Harris and C. J. Lindblad

This volume is directed mainly to grain loss situations in developing countries.

Determination of losses to food crops requires careful blending of the concepts and procedures of several sciences while each is given its necessarily detailed attention. Nowhere is this more true than in dealing with postharvest losses to grain. Information gathering ranges from A to Z, and at the outset emphasis needs to be given to the cultural-social aspects discussed in Chapter III.

While many of the methods contained in the manual relate to the evaluation of damage caused by a single organism or mechanical effect, such selective attacks rarely occur in nature. Interactions between major causes of losses must be expected.

A basic concept of this manual is that it be applied in its entirety. Care needs to be taken that personal, national, economic, cultural, and other biases do not generate unwarranted project plans or conclusions. To illustrate, large influential farmers may want technologies developed to suit their own needs which may be completely inappropriate for small farmers whose grain-handling systems are less mechanized or capital intensive, grain storage scientists may want to continue in their own research area to the exclusion of other equally important areas, national governments may favor one political region or group over another, or international development agencies may have their own priorities.

There are many ways to produce a list of intervention points. Consideration could be given to technological improvements that would both cost the least and prevent the greatest amount of grain losses to the benefit of the entire country as a whole. However, political, economic, and social priorities need to be taken into account in locating and identifying intervention points. What is technologically ideal may be very different from what is practical and feasible within the actual social, economic, and political environment. A balancing of technical and social sciences is essential in assessing and reducing grain losses.

For the purposes of identifying loss points which are critical and amenable to reduction, this manual uses the pipeline concept to describe the location and flow of grains. In this way, losses can be viewed individually and in perspective; however, the pipeline concept is not limited to technical or physical factors. Social realities come into play and perspective is required to both understand those attendant social influences and to prevent them from being blindly introduced as unrecognized bias. The pipeline approach weighs individual loss points in relative magnitude. Combined with consideration of social realities which influence amenability to in-depth assessment and loss reduction, the pipeline concept serves to 1) identify critical loss points for in-depth assessment and 2) provide a basis for development of improved technologies for postharvest loss reduction.

The influence of personal judgment, and therefore bias, cannot be avoided though the investigator or official may be unaware of its role. The investigator must also constantly guard against yielding to pressures based on unsubstantiated assumptions. An example of the consequences of this kind of oversight is seen in the countless huge, empty, and decaying grain bins installed across the developing world under incorrect assumptions. They serve to demonstrate that what is feasible in one situation will not necessarily be successful in another.

The compilers of this manual have operated under the well-reasoned opinion based on some practical experience that interventions to reduce grain loss are often best channeled to the farmer/producer. There are a number of reasons for this alignment. One technical reason is that the best form of loss reduction is early prevention — grain which is in good condition will deteriorate more slowly than grain which is, for example, already infested with insects or poorly dried. Following that logic, to assure good quality food grain throughout the pipeline, it seems practical and desirable to have it enter the pipeline under optimal harvesting, drying, and storage conditions. Another factor is that, in developing countries, much of the grain is stored and consumed in the rural areas, in large part by farm families.

A loss assessment study that does not have built into it the strong possibility and intention of benefiting the situation under study is of no consequence. *The purpose of loss assessment is effective and expeditious loss reduction.* Loss assessment need not and should not be a largely academic exercise.

Loss-causing damage may not divide into neat, exclusive categories. Moldy kernels may be insect infested and vice versa. Insects can cause shattering, and shattered kernels more readily support certain insects. Bits and pieces lost through holes in bags or in processing may have been produced by too rapid drying. These and other situations are more the normal than the exception and need to be duly noted and judgment applied in interpreting data.

Certain concepts are dealt with in only one section of the manual though they have applications throughout many facets of loss assessment and reduction. For example, while the subject of economics is in a separate section, it has applications throughout the manual. It bears on sampling and how, when, and where the samples are taken. It bears on the selection of study situations and how they impinge on each other, and it relates to cultural factors. Similarly, cultural factors are dealt with in a separate section though their implications are also pervasive as they bear on sampling, analyses, and the whole problem of functioning in a system without undesirably changing or destroying it.

Early in the preparation of this first edition, an attempt was made to prepare a manual that could be used by trained and untrained workers alike. This proved to be impossible. The ideal of writing for those without any background in grain storage, biology-entomology, food marketing, or the socioeconomic sciences was attempted and abandoned as impractical. The material is, therefore, prepared for people with at least some pertinent experiential or academic background.

One of the important matters not covered in this manual is the matter of mold toxins. This does not downgrade the seriousness of the mycotoxin problem. Important as the problem is, this volume is concerned with measuring losses of stomach-filling grain, not whether its nutritional value has been reduced. While strongly noting that food contaminated with mold toxins is to be avoided, as regards mold-caused losses, this manual deals only with such losses of grain actually discarded for human food because of the presence of mycotoxins.

II. TERMS OF REFERENCE

A. Definitions

K. L. Harris and C. J. Lindblad

This manual deals with food grains, cereals, and pulses and the word "grain" is broadly used to include all of these. It deals exclusively with the loss of food from the food chain and largely follows the definitions of Bourne (1). In it, a working definition of the term "postharvest food loss" is set forth as given below:

"POST HARVEST" means after separation from the medium and site of immediate growth or production of the food.

Post harvest begins when the process of collecting or separating food of edible quality from its site of immediate production has been completed. The food need not be removed any great distance from the harvest site, but it must be separated from the medium that produced it by a deliberate human act with the intention of starting it on its way to the table.

It does not include steps between cooking and eating as covered by Bourne and agrees with Bourne to "not cover inefficiencies in human metabolism and utilization of the food." In this manual, however, the pathway ends when the food grain or the food prepared from the grain, or both, reaches the point where it is to be finally prepared (cooked) for consumption.

Three periods of time may be identified during which food may be lost, and each period has its characteristic problems, and means of overcoming these problems.

a. *Preharvest* are losses that occur before the process of harvesting begins, for example, losses in a growing crop due to insects, weeds and rusts.

b. *Harvest* losses occur between the onset and completion of the process of harvesting, for example, losses due to shattering during harvest of grain.

c. *Post harvest* losses occur between the completion of harvest and the moment of human consumption.

Postharvest intermixes in varying degrees with portions of the maturing-drying-processing period and often no sharp distinction can be made. Thus, maize held in the field for drying is also maize held for storage and use. This manual does not imply that any artificial sharp distinction must be made.

Harvest and post harvest losses are sometimes combined into a single loss because there are some elements of common concern between them. A suitable descriptive term for these combined activities would be "post production losses". The following schematic representation shows the relationship among the various types of food losses:

1. Preharvest
2. Harvest   } Post Production
3. Post Harvest

In addition to Bourne's postharvest grain, this manual includes the ripe crop remaining in the field, whether standing in its original position or not, for further drying or holding, or both, until it is brought in or removed from the growing position, eg, maize drying/storage in much of Latin America.

"FOOD" means weight of wholesome edible material that would normally be consumed by humans, measured on a moisture-free basis.

Inedible portions such as hulls, stalks, [and] leaves . . . are not food. . . . Feed (intended for consumption by animals) is not food [unless specifically of interest to the individual assessment exercise].

The method of measuring the quantity of food in the post harvest chain should be on the basis of weight expressed on a moisture-free basis. There will be times when information on losses in nutritional units and economic losses will also be needed but these should not be the prime means of measuring post harvest food losses.

"GRAIN LOSS," as used in this manual, concerns the loss in weight of food that would have been eaten had it remained in the food pipeline.

"LOSS" means any change in the availability, edibility, wholesomeness or quality of the food that prevents it from being consumed by people.

Food losses may be direct or indirect. A direct loss is disappearance of food by spillage, or consumption by [insects], rodents, [and] birds. An indirect loss is the lowering of quality to the point where people refuse to eat it.

This definition is a people-centered definition. "Food" means those commodities that people normally eat and excludes the commodities that people do not normally eat. If the food is consumed by people it is not lost; if it is not consumed by people for any reason at all then it is considered a post harvest food loss.

Food losses are, at times, simply as they are locally defined or as they locally occur. For example, grain which is discarded because of discoloration is a loss.

Processing losses occur when edible portions of food are removed from food channels by the process or by spillage or breakage from the process. Rice hulls are inedible. Their removal does not constitute a loss. Rice pieces diverted from the food-chain are a loss. Rice bran is edible to some, inedible to others. The handling of each similar situation needs to be clearly defined as it occurs. Corn cobs or cores are not a loss. The corn seedcoat is removed in making corn grits. It is not removed in making many other foods. How it is handled needs to be defined in each appropriate instance.

Where quality deterioration results in a loss in weight or in the food not being eaten at all, eg, rejected in the marketplace, the rejected food is a loss. In this volume, quality is a consideration only as it relates to loss in weight of food, but how it is handled needs to be defined appropriately in each instance.

The term ''insects'' includes true insects (six-legged arthopods) and grain-damaging mites.

Microbiological losses and microbial losses are used interchangeably to refer to losses caused by molds, yeasts, and bacteria.

Literature Cited

1. BOURNE, M. C. Post harvest food losses — the neglected dimension in increasing the world food supply. Cornell International Agriculture Mimeograph 53 (1977).

CHAPTER II

B. Planning: An Overview for Project Administrators

K. L. Harris

Determining agricultural losses involves many disciplines and goes to the heart of established cultural patterns. Administrators need to recognize the complexities of what they have to deal with and understand that unless defects in planning and implementation are overcome, the results will be jeopardized. While this is an obvious platitude, it is of special importance here since the nature and quality of the operation can set the stage for the nature and quality of other programs that may follow in the technical and lay community.

Without attempting to set forth an administrative manual, the following details are to be noted:

1. Project planning, depending on circumstances, may require inputs from, for example, agricultural economics, agricultural engineering, agricultural extension, administration, anthropology, biology-zoology, cultivators/grain owners, education, entomology, food marketing, grain storage science, microbiology, political science, rural sociology, and statistics.

2. Revealing the status of the food grain supply may be a delicate matter that impinges on matters of national and international security, as well as on local, national, and international commodity markets and on foreign exchange balances.

3. One needs to be aware of social factors; special village allegiances and requirements; the role of women, the family, and other groups; and whether information is best collected by lower-status field-workers, peers, higher-ranking individuals, etc.

4. Logistic requirements are imposed by terrain, delineated and undelineated boundaries; presence or absence of containers, scales, meters, transport; local customs and work patterns; and training requirements and capabilities.

5. Assessment work needs to be understood in terms of cultural factors: local vames and definitions and local social and agricultural systems.

6. The assessment must relate to local needs — individual, national, and all in-between.

7. One should be aware of the interrelations between postharvest losses and preharvest.

Basic survey operations, schedules, and plans are set forth in Table I and Fig. 2. The time needed for such a survey will obviously depend on the size of the country and accessibility of the sampling areas, but the decision on the selection of farmers must take place before any final work begins so that sampling visits can start immediately after harvest or any other start-up time. Modifications to the sampling pattern may be made in the case of crop failures or similar unavoidable circumstances.

The nature of the operation — and planning for the operation — will depend primarily on the factors that are to be investigated and how they are to be investigated. This is the subject of this manual.

This manual deals with 1) obtaining a planning overview of grain movements, the grain pipeline, 2) determining what portions of the pipeline should

TABLE I

Basic Plan of Operation

Stage	Timing Weeks		Activity	Personnel[a]
Preharvest	6	1	Familiarization with local agricultural structure and geography	CO
	2	2	Preliminary survey for choice of sampling areas	CO, ES
	2	3	Fact-finding visit to chosen sampling areas for information on storage practices to identify strata and select appropriate method of obtaining farmers	CO, ES
Harvest[b]	up to 4	4	(If required, construction of experimental stores)	CO + laborers
	2	5	Initial visit to selected farmers to obtain basic information and baseline samples (also purchase grain for experimental silos)	CO, Exp, ES, LA
Postharvest[b]	1-3	6	Examination of baseline samples in laboratory and check on proposed methodology	Exp, LA
	1 per month	7	Monthly sampling visits to selected farmers to collect samples and record consumption patterns	LA, ES
	1 per month	8	Laboratory examination of field samples (and experimental samples)	LA
	7	9	(If required, brief questionnaire survey of other farmers to confirm storage pattern)	CO, ES
	2	10	End-of-season visit to selected farmers to check consumption and thank for cooperation	CO, Exp, ES
Next Harvest	4	11	Analysis of results in terms of loss per sample and integration with consumption pattern	Exp
		12	Preparation of report	CO, Exp

[a]CO = Country project officer; ES = extension staff; Exp = expert TSPC; and LA = laboratory assistant.
[b]Drying, processing, bulking, etc.

Adapted from: Tropical Products Institute, Tropical Stored Products Centre, Slough, England.

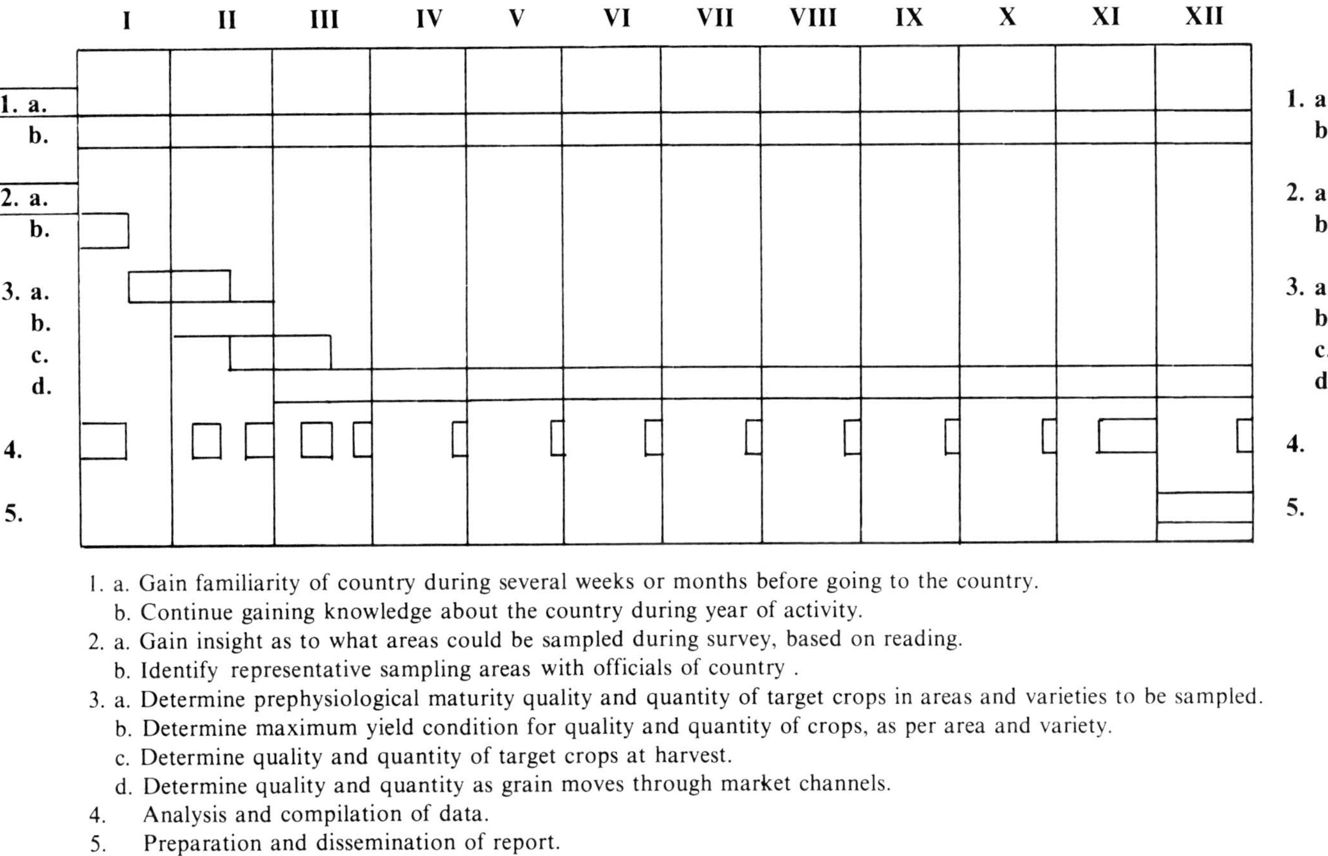

Fig. 2. Time allocation during a loss survey. (Adapted from T. A. Granovsky, Kansas State University.)

1. a. Gain familiarity of country during several weeks or months before going to the country.
b. Continue gaining knowledge about the country during year of activity.
2. a. Gain insight as to what areas could be sampled during survey, based on reading.
b. Identify representative sampling areas with officials of country .
3. a. Determine prephysiological maturity quality and quantity of target crops in areas and varieties to be sampled.
b. Determine maximum yield condition for quality and quantity of crops, as per area and variety.
c. Determine quality and quantity of target crops at harvest.
d. Determine quality and quantity as grain moves through market channels.
4. Analysis and compilation of data.
5. Preparation and dissemination of report.

be further investigated both because of the size and nature of the losses and their feasibility for reduction, and 3) conducting the detailed investigations.

This manual also stresses the use of existing in-country data on what grains are produced in what quantities in what regions and consumption patterns.

CHAPTER II

C. An Overview of the Postharvest System: The Food Grain Supply Pipeline (Determining the Interrelationship and Relative Magnitude of Losses)

K. L. Harris, W. J. Hoover, C. J. Lindblad, and H. Pfost

The flow of grain from its sources, ie, the farm field or import docks, to the eventual consumer is depicted for the purposes of this manual as a pipeline with many possible interconnecting pipes and reservoirs. Losses, or leaks, can occur along the entire pipeline — during harvesting, drying, transport, storage, and processing. As presented in the Preface, the purpose of viewing the food grain supply system as a pipeline is to assign individual loss points (eg, on-farm losses) relative importance in terms of loss in other parts of the grain pipeline (eg, transport or warehousing losses). This relative perspective is necessary to see the importance of the total amount of grain actually lost in any given point as opposed to the percentage of grain lost which passes through that point. Failure to obtain such a perspective has resulted in overly high and low loss figures arrived at by extrapolating from observed losses at specific loss points without putting those losses into the perspective of the grain moving through the total system.

This failure and the need to obtain an overview often apply to expatriates and others entering a system for the first time.

One needs to use all possible local information to determine how and when the grain moves from harvest to consumer, routes for movement and holding patterns, and where and how processing is accomplished. Most of this information is known locally.

Grain does not move in a straight line and uniform sequence from producer to consumer. Harvested grain can be specially dried and otherwise treated to go into special household use; some into an even more special seed-grain storage. This grain may remain there or move out for food or trade under special conditions influenced by factors such as family, weather, or government. It may even be replaced by other local or imported grains. A portion of the harvest may be held for short-term storage, a part for long-term storage, and the rest sold or otherwise traded off the farm.

All of these factors, and more, need to be kept in mind in determining where and what should be tested.

Delineation of the test sites involves looking closely at general loss situations and careful on-site evaluations of specific individual sites. Selection of "amenable" sites (villages, cultivators, markets, transit systems, warehouses) requires incorporation of many factors. Accessibility must be balanced against the location being atypical because of proximity to outside influences. Traditionalism must be balanced against the need for outsiders to be accepted into the delineated area. Language can be a key barrier, and an absence of direct or completely competent and trusted lines of communication is unacceptable for loss survey teams. Sex roles must be considered as to who really does the harvesting, threshing/cleaning, storing, and marketing of the grain. All pa-

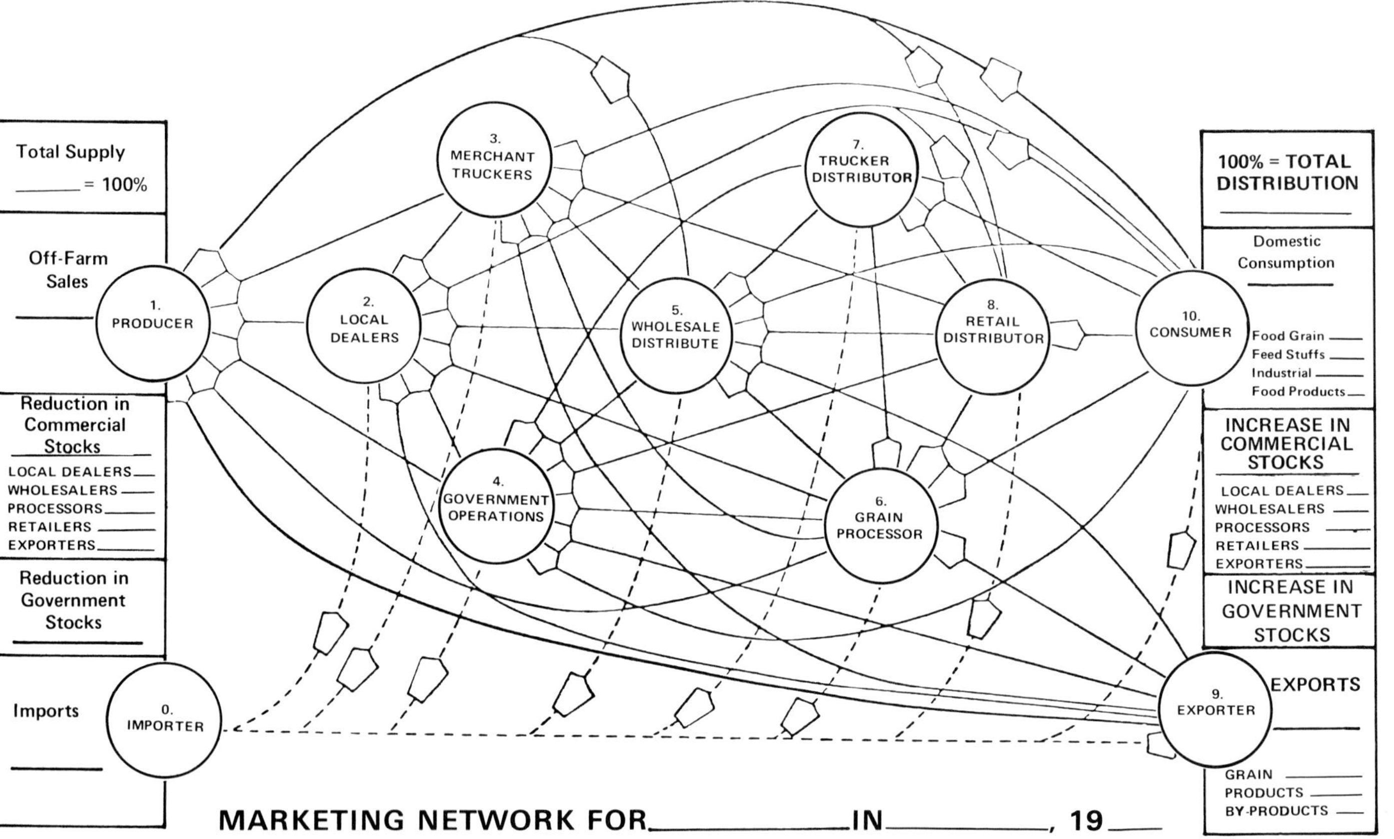

Fig. 3. Grain marketing flow patterns. (From Department of Grain Science and Industry, Kansas State University.)

rameters need to be considered, and should cover the entire social, cultural, physical, commercial, and political setting.

Even the simple village market has flowing through it all these effects, and more, so that if there were to be a single measurement it would, in reality, consist of measurements of many factors, each weighted as to volume.

Knowledge of actual high-loss and low-loss situations is required in determining the need for, location of, and types of interventions. However, inordinately high- and low-loss situations must be put into perspective rather than giving them overemphasis as has been the case in some instances.

To further illustrate, out-of-condition grain held by market speculators may suffer very high losses, say 30%. Taken by itself, this level of loss might identify grain speculators as a critical focus for improved storage technology intervention. However, if in fact only 5% of the total grain supply is ever handled by such speculators who specialize in out-of-condition grain, the real value of the total losses at this speculator level becomes 30 × 5%, or 1.5%

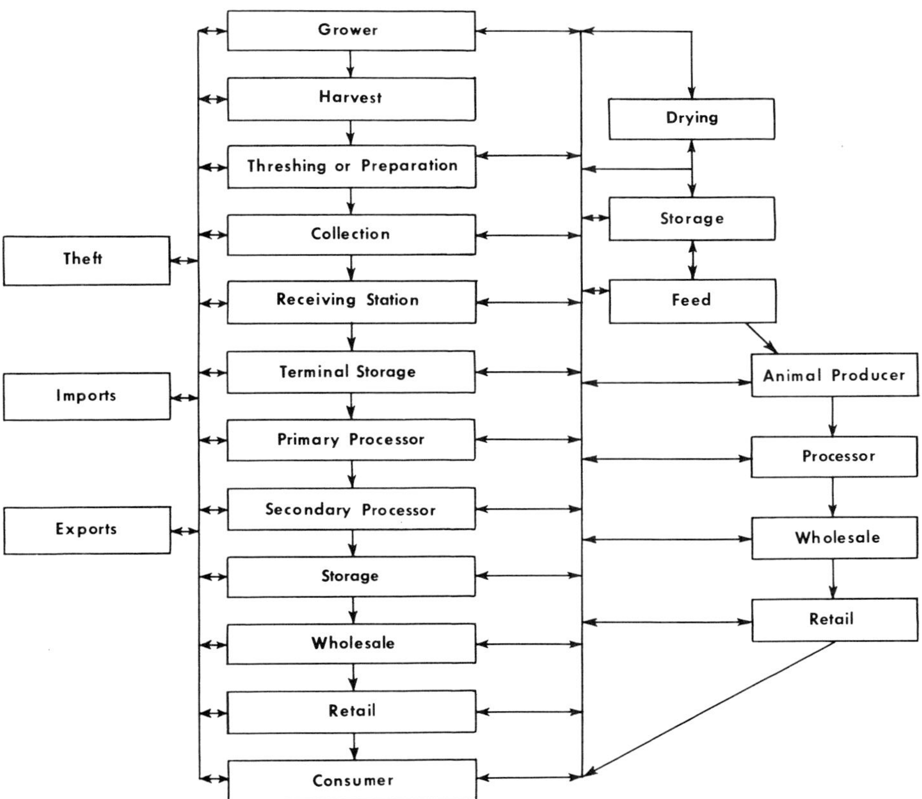

Fig. 4. Grain use flow patterns.

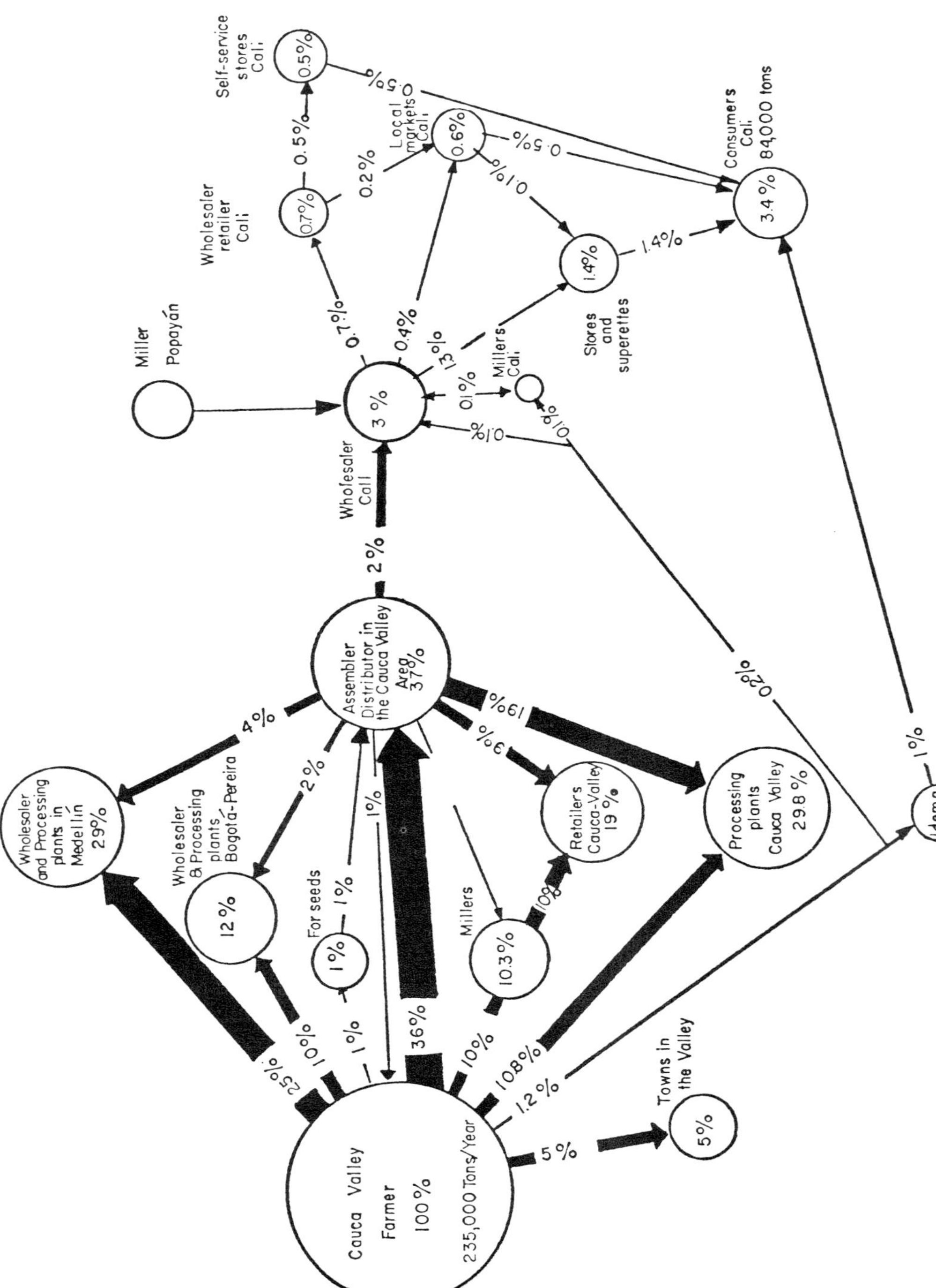

Fig. 5. Channel map of corn distribution, Cauca Valley area, 1968. (From K. L. Turk, ed. In: Some issues emerging from recent breakthroughs in food production. Cornell University, Ithaca, NY [1971].)

rather than 30% of the total grain supply.

A useful investigation of postharvest grain losses requires detailed knowledge of the entire postharvest food grain supply pipeline. Figures 3 and 4 are two representations of supply pipelines. Figure 3 emphasizes marketing patterns; Fig. 4 emphasizes the processing flow through to the consumer. At any one point, grain or grain products may move out of one pipeline, around several intervening steps, and re-enter further along in the sequence. Similarly, movement occurs in both directions. Grain gleaned from the field or from spillage on a farm or in a rural market can go immediately to a consumer or may be bartered back into a trade channel. What might be loss to a farmer by spillage at a local market, or to a transport company, may in reality be a mode of payment for services rendered at an otherwise unacceptably low pay scale.

In each country, district, or community area, there exists a marketing system for food grains. It is imperative that the flow of grain through the various facets of this marketing system be quantified so as to establish priority points for observance and measurement of losses, and to subsequently focus attention on loss prevention programs. Figure 5 shows a quantified flow in which different grains and oilseeds follow different routes.

Moreover, different parts of the pipeline have different flow rates. While a particular grain may be in a storage chamber for some time, it may be in a milling process for a very short time. The types of losses at those two locations are different; one is a loss which increases over time, and the other is probably a one-time loss due to such things as poor physical handling, equipment, or packaging.

To follow the pipeline analogy, the two types of losses occur in the reservoirs and in the pipes. Once grain has passed through a leaky pipe (eg, a poorly adjusted grinder), it is not subject to that particular loss any longer. However, grain in a holding reservoir (eg, a rodent-infested bin) is subject to those losses for as long as it remains there. Loss assessment methods and calculations for the two types of losses can be quite different.

This, of course, complicates the task of assessing losses. Separate measurements are required for the different types of losses that occur due to mishandling or poor equipment settings, in addition to the biological deterioration caused by insects, rodents, or moisture or other climatic conditions. Sampling, tracing, eventual utilization, and testing of overall losses really entail making and evaluating individual components in a system and calculating their overall effects. Moreover, since effective loss reduction interventions need to be directed to the reduction of specific leaks, it is the individual loss figures that need to be evaluated, not overall national figures.

Note: With the acknowledged limitation of development resources and perhaps even greater limitation of available, trained personnel, the pipeline concept is an approach that is recommended as a means of quickly and inexpensively focusing on significant losses in the overall system. It is also an effective procedure for effective resource allocation.

There is every reason to believe that the presence of a survey in the system will itself affect the system and the results of the survey. This will not be discussed other than to note that economic, cultural, and political factors governing the flow and treatment of grain can be expected to respond to the survey itself, thus partially skewing the results.

CHAPTER II

D. Preliminary Examination of Specific Problem Points and Making On-Site Rapid Appraisals

G. G. Corbett, K. L. Harris, H. Kaufmann, and C. J. Lindblad

Two of the most critical aspects of postharvest grain loss methodology are the need to not attempt more than is feasible, and to rapidly seek and identify for investigation major loss situations that seem both amenable to study and responsive to improvement through practical interventions. By using a pattern that has found almost universal application by expatriates from international and national agencies whether dealing with the most primitive situations or the most sophisticated, this first appraisal has become accepted as a 30-day exercise. However, 30 days may be too little or too much time, although this will only be determined by the complexity of the system and the nature of the questions being asked.

As with any investigation, some early judgment is made that the work is needed and that there is a reasonable likelihood that useful results will be obtained. After that there is a need to work with local officials in a preliminary fact-finding canvass of the situation that goes into the entire nature of the grain pipeline, as explained earlier, and then into individual problems and their projected solutions. This includes all of the grain movement logistics, personnel, political and cultural ramifications, etc., that will be called into, or will force themselves into, the final study.

It would be well for this preliminary canvass to proceed solely as a prelude to a larger study, but such will not always be the case when immediate developmental decisions must be made before detailed information can be made available.

Interrelated aspects will proceed together during the 30- to 60-day preliminaries:

Assessment exercises may be undertaken by expatriates to determine losses, while locals seek to determine how to reduce the losses.

One task is the probing for specific problem points; the other is the job of making rapid on-site appraisals.

One looks ahead to a more definitive investigation; the other comes to on-site loss and intervention judgments within the rapid-assessment time span.

In one case we are developing a strategy to conduct a survey; in the other, the survey and loss reduction efforts may be rapidly under way.

Preliminary Examination of Specific Problem Points

An initial survey is needed to determine what the problem is and what has to be done. In the initial survey the best possible information available should be used to ascertain the order of magnitude of the losses in the whole postharvest system and to identify the major points and causes of losses. As the loss figures are evaluated and observed to be accurate or inaccurate, they may serve as data to evaluate the local system. It is important to obtain information from people who are knowledgeable of the factors being assessed as well as from voluble proponents of biased or special interest positions. Already available reliable information, or lack of it, will help to decide the depth and focus of the preliminary mission.

The key element is to identify those problem points that can be adequately isolated, are likely to yield useful information, and are amenable to study and loss reduction intervention.

Few locations in the grain pipeline will be neatly packaged, single-entrance, single-exit, one-measurement situations. It may be necessary to make measurements over a period of time, to identify the points at which important losses are occurring, and to make an estimate from the data and evidence available of the order of magnitude of these losses. After such a survey (which will probably reveal the need for longer-term assessment of losses), it will be possible to define immediate, as well as longer-term actions. At the same time, the cost/benefit implications for both the operators concerned and the country as a whole must be considered.

The composition of the 30- to 60-day preliminary investigation mission will vary according to the complexity of the grain industry and the local information and expertise available. At least a grain marketing economist and a grain storage specialist (entomologist-biologist) should be included plus a processing specialist if it is anticipated that processing losses at village or industrial level are important.

Members of the preliminary mission *must have experience in the organization and operation of the grain industry in developing countries.* The social skills acquired by direct experience are invaluable and essential for the judgments which must be made during the preliminary survey. As experience is so critical here, interns would usefully be included in the mission; however, large missions (more than four) are often hard to accommodate within traditional social structures.

The mission will:

1. Map the pipeline using available government statistics and other inputs from key informants.

2. Conduct an initial survey of the postharvest grain sysem to establish who is handling, storing, transporting, and marketing the harvested crop; what part of the crop is handled and stored by each operator, and for how long, including farm storage for self-consumption purposes; and the condition of handling, storing, and processing.

3. Review all available data on quantitative and qualitative losses occurring in the system(s) and identify the major causes and extent of loss.

4. Prepare an inventory of available storage, transport, marketing, and processing facilities and assess their adequacy in capacity, design, and condition.

5. Review the present activities being undertaken to reduce postharvest losses and list the resources available for these activities from both internal and external sources.

6. Design a phased action program to investigate or implement under the project terms of reference.

In conducting the preliminary study, remember that grain losses occur in situations that cause or allow them to occur, and as the losses occur, evidence is left of what has and is happening and what will probably continue to happen.

There are many clues to both general and individual aspects of grain losses that can be disclosed by the rapid assessment of a situation. Knowing that key elements in insect depredations are moisture, temperature, numbers and kinds of insects, length of storage, storage sanitation, and insecticide use and other control practices, one can keep the presence or absence of these factors in mind and come to some general or specific conclusions based on known scientific principles. Estimates of 30% losses to maize stored for several months under humid tropical conditions may be quite reasonable. The same figure when applied to a cold, dry climate or to grain used up in three months may be unreasonable.

Many farmers are well aware of these factors. Out-of-condition grain is often passed along to the local market or government agency. Grain for long-term storage may be dried, put in better storage, or treated wth a protectant. Loss-prone varieties may be used first or sold off the farm.

Some conclusions will be fairly straightforward. For example, if grain goes into bagged and naturally aerated storage that has evolved within the culture, reasonably good storage quality may occur. If the same high moisture grain goes into sophisticated silo storage without the necessary sophisticated drying, there will be a high potential for loss. Poor sanitation, insects, molds, leaking roofs, rats, uncleaned bins and bags, high atmospheric humidity, and extreme temperature variations all affect grain losses.

Generally, when insect damage is very difficult to find, the weight losses due to insects are also negligible. One may know what a 25% loss in maize looks like in one region and carry this mental picture to other regions and other situations. The significance of frass, of extensive moth webbing, of adult or larval insects may be so well known that they automatically lock into a fairly accurate judgment — a judgment that may well be sufficient for the experienced person to come to a general conclusion on the extent of the losses themselves. This, in addition to contributing to a decision on whether a situation should be tested or surveyed in depth, may be as much as the situation warrants, especially if the losses are estimated at around 5%. At this low level, even an in-depth assessment based on currently known sampling procedures would probably be subject to an error as large as the loss.

In short, it is possible to do an overall appraisal based on an expert evaluation of the system with attention to pertinent parts of the harvest-to-consumption flow or patterns, and to such loss-inducing and loss-reducing factors as:

1. Moisture
2. Temperatures
3. Insects, rodents, birds (kinds, numbers, association with the grain)
4. Length of holding
5. Local quality and quantity controls
6. Types of bins and other holding vessels
7. Sanitation-insanitation
8. Trading quality factors
9. Use and nonuse of pesticides

   10. Evidence and nonevidence of grain damage; kinds and amounts
      a. Frass and webbing
      b. Exit holes
      c. Darkened (rotten) kernels
      d. Degermed kernels
   11. Mechanical loss factors
   12. Location in the harvest-to-use pattern

The need to apply the physical loss parameters and to know what stimulates or retards losses cannot be overemphasized. Many unreasonable guesstimates would have been avoided if more attention had been paid to such criteria. Of course, these same criteria will provide an operational arena for in-depth assessment and loss reduction.

Finally, one needs to remember that just as losses do not occur in a vacuum, neither do loss assessments, and one should expect the presence of a survey — with or without an overt attempt to make improvements — to induce changes.

III. SOCIAL AND CULTURAL GUIDELINES

The overall aim of this chapter is to introduce some of the complex cultural-social-anthropological factors to postharvest grain loss assessment/intervention activities. The message is made up of a variety of signals that pass in both directions: from the situation being investigated to the investigator and from the investigator to the situation. It is a dynamic process.

In grain loss assessments the need is to find out what the situation was or is. The investigator wants to affect the milieu as little as possible while he assesses it. Thus he needs to be in tune with what is happening so that the assessment will be an assessment of what he sets out to assess — not of what his presence is bringing about.

This chapter is a result of many discussions, not only with Allan Griff and Conrad Reining, but with many others. Griff, Reining, Harris, and Lindblad, together with Edna Loose and Maryanne Dulansey, examined, analyzed, and reasoned the subject many times together. What has resulted is the foundation statement of Part A and the evocative of Part B. Part A is self-explanatory. Part B is purposely set forth so as to leave the assessor with many questions into his own investigations.

A. The Fact-Gathering Milieu

Allan L. Griff

It seems obvious that planners and field-workers of grain recovery programs should be familiar with the social and cultural background of the places where they are working. But far too often this knowledge is insufficient and incorrect, and the result can be error and waste. Cultural awareness is no guarantee of success, but it can help.

This chapter is but a brief outline of how culture operates, and its place in the early stages of planning a program. It will raise many questions. It may slow down some projects until adequate understanding of the people is achieved. It may improve communications enough to get some projects off a comfortable and self-perpetuating dead-center. But if one is committed to tangible results rather than just good appearances and completed missions, culture cannot be ignored — rather, it must be understood. Culture is on our side. Few want grain losses, but only a good understanding of the roles of social and economic behavior of the people involved (ie, the culture) can make this a contributory factor and not an adversary.

Culture is not Static Tradition

First, we must erase the stereotype view of culture as stubborn adherence to tradition and resistance to change. All cultures contain the seeds of change as well as the inertia to resist change. This is the basis of cultural evolution. Changes can and must occur for a society to survive, but they must be opposed and tested to ensure that they achieve their aim, that the gains are worth the losses, and that change does not occur so fast that the people cannot adapt to it.

In this light, we should realize that what we think is good change, or even what a country's leaders think is good, is never 100% good. There is a price to pay for all change, and much resistance arises because the price is too high for some or just cannot be paid without excess hardship, despite apparent longer-term value.

Some people in some countries are used to a logical, scientific sequence of cause and effect and can thus predict the future, more or less. This enables them to confidently invest time, labor, and money in the future. It gives a sense of control.

But in many developing societies, the people have little control and they know it. Their plans have been thwarted by natural catastrophe, or by magic, or by the will of forces distant and far more powerful (including both gods and central governments). Given the crawling pace of development among the world's rural poor, we cannot blame them for being a little skeptical about proposed changes. This is not necessarily blind tradition. It may be healthy and justified caution.

And stability itself has a positive value in all societies as it reinforces behavior by promising future returns for today's behavior patterns. Without stability, people lose the incentive to keep past social values, as future outcome can no longer be predicted. The result is an explosive proliferation of values (witness America and Europe today) and a disincentive to plan for the future at all.

Evolved Versus Imposed Change

Many cultural changes have been imposed on people, often suddenly, with remarkable results attesting to the equally remarkable adaptability and resiliency of people. Conquerors and rebels have imposed languages, religions, food habits, and codes of law on other people since prehistoric times. They have often also brought innovations that were eagerly adopted by the local people, such as the gun and horse among American Indians, and baseball and hamburgers in Japan.

On the other extreme, some changes took many generations to evolve, perhaps because they were not very important or were not enhanced by political association, or perhaps the price to be paid for the benefits was high. Where agricultural innovations were concerned, the risk was often simply too great. Some people lived and still live too precariously to experiment even if the idea looks promising.

Development strategists today are caught in the middle. They do not want to impose, yet cannot wait for evolution to do the job unaided. So we have derived an intermediate form of "coaxed" change, in which we decide beforehand what change is desired. People do indeed want to better their lot, but may be convinced that such efforts are futile and may be too polite or too scared to tell us so, or may not even realize why they resist. Therefore, it is a good idea to look at the recent history of the subject community to see how changes take place in that community.

Study the Past

Every group has its own ways of change. They usually are those that minimally disrupt the effective social order, and are also in tune with the popular trends as evidenced by past change. Thus, both present and past — in this case, related to the economic and interpersonal structure of food storage and use — must be appreciated to see what might work and what might not. To this end, the following questions will be useful:

1. Has the community made technological or agricultural changes in the recent past? If so, through what channels were the changes introduced? Were there models to copy? Key people whose support and influence were critical? Economic or other incentives? Were the changes mainly imposed, coaxed, or naturally evolved? Are the changes now an irreversible, integral part of the culture, or are they artificially supported by current leadership and likely to revert to original status if the support were removed? (The potential permanence of a change is as much a measure of success as the change itself.)

2. Have any change attempts failed in the recent past? What were their histories and apparent reasons for failure?

With regard to the "who" questions, the models are particularly important and simply any model will not do. Certain people will be followed, others rejected, still others ignored. The one who acts first may not be the real leader; he may be marginal with nothing to lose by trying or he may be acting under the influence or command of others. The area of influence is also important — a man who can command respect and honor among civil servants may not count for much among the farmers, or an older leader may be resented by the young, and vice versa. It pays to learn local history to see how things got done before, and for expatriate workers it is certainly an error to assume one's own national patterns of power and influence will apply.

It is also dangerous to believe everything we are told. Observation of attitude and even tone of voice may be as important as the actual words said. Double-checking critical statements is essential; relying on one or two data points is as inadequate in social science as it is in physical science.

How Do You "Learn" a Culture?

The most obvious answer is time — implying that people who have spent years in a group become expert observers of that group. This is not always true. Of course, time is necessary, but a competent observer must also know how to observe, must be himself/herself relatively free from familial or political involvement that might affect observations, and must be articulate enough to transmit them to others.

In dealing with local sources of information, all individuals are not equal. Some are "balloons" — innovators who are free to change and the first to do so, and some are "anchors" — social-role conservatives who provide and represent stability. Local landowners and similar elites are often in this class, while their children may well be balloons as with a relatively secure future they can afford to be different. This balloon-anchor continuum is a convenient way to characterize local contacts and ultimately to ensure that one's information does not all come from one type.

Just as people's responses depend on their individual characters, they also often depend on how they view their questioners. Association with the local government or a donor agency may be helpful in some cases and a handicap in others, and a strong personality may turn a respondent in many directions. As an agent of change, an investigator must not imagine himself free from bias either. Attitudes toward development and efficiency are hardly universal. But he can try to stand back and put his own values aside for a while, at least while working, to enable him to learn what makes a host community tick. This will be necessary to work within it to achieve the goals he has accepted for the project or, when that is impossible, to get out gracefully.

Talking to natives or experienced foreigners may be the next best thing to living for years in a place, but these are not the only alternatives. For some people, it is easier and better to watch and listen to others without asking questions. It is certainly less intrusive. Often, a conversation about events seemingly unrelated to grains and farming will reveal ideas and attitudes which affect the proposed actions. Economic insecurities, anxiety about family nutrition, worry about too much centralized control, and local labor problems are examples of things worth listening to. Reading local newspapers and attending local public functions where appropriate are useful techniques; beware, though, of being inadvertently classified with a political party or social class that is linked with the newspaper or the function. In any case, keeping eyes open, and perhaps keeping a diary of observations, will pay off. And if your function and aim become well known, you will receive much useful information.

In some groups, the very existence of a foreigner implies change and is a threat to some and an object of economic courtship to others. It is hard for foreign experts to avoid getting tangled in political games; if we have money to spend or control, we are obvious objects of interest and concern. In some of the more cosmopolitan places, however, where agricultural development and extension work is commonplace, a new face is more easily accepted. Unfortunately, the very places where acceptance is easier are also those with more complex and intricate social and economic relationships, so the job is proportionately more complex.

Culture or Cultures?

It is convenient but rare to find a homogeneous community with similar beliefs and behavior. More often there is a continuum of behavior from traditional to daring, and sometimes a sharp age distinction, separating the younger people who grew up after World War II in an atmosphere of independence and international communication, from the older generation for whom tomorrow was expected to be the same as today or yesterday. Sometimes the split is between urban and rural, or factory workers and farm workers, or on racial or religious lines, and, of course, there may be more than two groups involved.

The careful observer, then, will not automatically assume "one culture" but look for signs of pluralism that will help him to identify, classify, and eventually understand the different attitudes and behaviors of different people.

Advance Preparation

Much can be learned before ever setting foot on the location to be studied or assisted. In almost every area of the world, hundreds of observers have already been there and, consequently, there are hundreds of books and articles telling about the people and their cultures, ranging in quality from useless to marvelous. Therefore, it is inexcusable not to study in advance.

Most field-workers get basic country information through their own agencies, the host governments, or their own government post descriptions. These are adequate if they are up-to-date and not too strongly aimed at visiting businessmen and officials who do not have much contact with the rural people. A more subtle problem is the definition of a country or region through the eyes of its own U.S./Europe-educated officials and managers. These people may ignore basic aspects of the culture because, with good intentions, they think they are useless blocks to progress.

Sources

More detailed cultural information is available and worthwhile. Some sources are:

1. The American Anthropological Association, which has a division concerned with agricultural development, with names and members keyed to regions and experience topics. Contact John Bennett, Washington University, St. Louis, Mo., or Iwao Ishino, Michigan State University, East Lansing, Mich.

2. The anthropology department of the nearest major university. In checking, you may find a student just back from a year's fieldwork there and eager to tell you what he knows, or a professor who is a recognized authority. Or the faculty may know who in other universities would know what you want, as this discipline is a complex and well-functioning information network in itself. (A word of warning: "anthropology" outside the English-speaking world sometimes is narrowly defined as study of physical characteristics and perhaps of primitive tribes. In these areas, study of the culture-linked aspects or agricultural behavior may be found in departments of sociology, ethnology, economics, or in agriculture itself.)

3. The Human Relations Area Files at Yale University, New Haven, Conn. which has cultural information on most of the world. You do not have to go to New Haven to use it, as many other universities have access.

4. A meeting of appropriate professionals, such as the American Anthropological Association which meets each year in November, with numerous speakers, and its subdivision on agriculture meets at that time as well. A related and useful organization is the Society for Applied Anthropology which meets in the spring of each year. Details on both groups are available from their common headquarters at 1703 New Hampshire Ave. N.W., Washington, DC 20009.

5. The Society for International Development, an organization of development professionals — economists, technical consultants, officials, and fieldworkers in aid organizations, and a few anthropologists. Most of the members have international field experience. There are chapters all over the United States and Europe as well as in some developing countries. The New York and Washington chapters are the largest and hold several meetings each month; the Washington group even has a rural development subdivision. For more information, contact the North American office, 1346 Connecticut Ave. N.W., Washington, DC 20036, or the world headquarters at Palazzo Civiletta del Lavoro, EUR, 00144 Rome, Italy.

Sources in Developing Countries

If you are already in the field, it may be difficult to reach many of the sources noted above. If there is time, you can write to them (offer to pay for Xerox, book, and airmail costs). But if you have to gather knowledge yourself, there are still a few things you can do.

If you have prepared a grain flow sheet (or pipeline chart) — a diagram showing the channels and amounts of grain as they move from farm to consumers — there will be certain key locations that control movement. Going to these places and watching who does what can be very useful, if it can be done without obvious intrusion. For example, watching who buys grain at a central market will yield information on purchase quantities, which in turn tells us about home storage. If no money changes hands, there may be a credit situation which controls purchase.

Watching harvesting and transportation of grains is also useful, and often possible in the role of technical expert. But it will help to learn who the workers are, who owns the vehicles or animals, what happens to spilled grain, and other such factors. The object is to understand the economic relations among the people and ultimately to understand the potential effects of any proposed changes.

Officials and local counterparts in a grain saving program are certainly available sources of information, but must be heard with caution. Some are farmers themselves, or have worked in the grain pipeline for years, but others may not really know how the majority of farmers and consumers behave; or they may not want to talk in detail about behavior that they consider old-fashioned or even embarrassing. We do not wish to imply that all or even most local officials are devious or misinformed; we warn only against uncritical acceptance of their descriptions without other indications or feelings that they are sensitive to and reporting what is going on around them.

There are many information sources in developing countries beyond the officials. Many countries have a strong awareness of their own cultures and have much published research. Appropriate university departments and libraries as well as government officials can be helpful.

It is often useful to look at people through the eyes of observant and articulate members of one's own culture. They can anticipate problems and reactions, and their advice should be sought. These might include anthropologists in the field, workers for volunteer organizations, or missionaries.

Key People

It is important to identify key people who can influence acceptance of changes, but it is also important to distinguish between apparent influence and real influence. Some people in important positions may really be servants of the position and cannot promote certain changes even if they wanted to. (This also is true in Europe and America.) Thus, personal and logical argument will be useless and may even embarrass the official who knows you are right, but is reluctant to explain why he must disagree.

Some positions of authority are temporary and others permanent, so it is important to know the system by which people come in and out of power. This can be quite complex — in some areas, for example, people move up both religious and political ladders, switching back and forth in a traditional pattern.

Many of these traditional systems are breaking down in the face of modern technology, communications, and other influences. Sometimes a foreign fieldworker finds himself/herself a symbol of change, with corresponding personal alignments and antagonisms, even before he ever says or does anything. This is a hard position to be in and some projects are doomed to failure or dormancy (a more polite and often more profitable alternative) no matter what the technical or economic merits of the proposed actions. Even if nothing can be done, it certainly is good to be aware of such situations and perhaps ask other colleagues about them on arrival, as part of initial briefing.

The Culture of Development

The development business has its culture, too, involving both foreign agents of change and local managers. Everyone has his/her own interests, and it is reasonable to expect people to act in their own interests.

It is often easy to blame inaction on a few individuals, or on one class of people or another, but development is not that simple. In reality, people of all classes will resist risk, even as they desire growth and improvement of their lot, if they sense the chance that their status might change for the worse.

From this need to minimize risk emerge the relations among government people, local businessmen and farmers, technical experts, and representatives of foreign and domestic money sources. These relations build, of course, on existing socioeconomic patterns, and are themselves dynamic, changing as needed to maintain development money input with minimal disruption.

In each location, this network is unique, and there can be no fixed guide to inform the newcomer, but a discerning fieldworker can easily see what is going on. Observe the social relations of the participants — who is invited by whom, who accepts and who can reject, who pays at lunches or dinners, what reciprocity is expected and what is given, who visits and who stays put, and who waits for whom at appointments. Watch language cues, too, such as the use of the familiar verb forms, first names or nicknames, and dialect or slang in direct conversation.

In any such network, some people are more free to act than others, and this degree of freedom should be noted for the people one must work with. In general, technical experts have more freedom (but less power) than political officials, young or old people more than middle-aged family heads, people from another area more than others with local family and business connections. These are guides, of course, and not rules, and there will be many exceptions.

In some places, there are long-standing patron-client relations which keep subsistence farmers in permanent debt and service, or else maintain them as low-paid farm workers. To the patrons, anything that may increase the economic power of their clients — even a grain use survey — may be seen as a threat to the current status, often already endangered by the communications revolution. Some patrons are very troubled by this; others do not care. They will all usually cooperate with government and change-agents, and many really do want their people to eat better if that were possible without disrupting the entire structure that they feel responsible for maintaining. In fact, where leaders are sufficiently secure as to be benevolent in deed as well as word, there is the greatest chance for successful change, as the leadership can then get things done.

A special problem is the self-perpetuating project which employs many people including international civil servants, is government-sanctioned and supported, and has no place to go if it succeeds. Thus, projects are kept in a state of incipient success to assure the flow of money and support, as well as the absence of disruptive change. Seldom is this a conscious conspiracy; more often it arises from the very nature of the situation.

Much of this is common knowledge among careful analysts of the development business. We include it here, though, because it may be useful for field-workers new to development, and also because the interface between field-workers and local officials is an area worth more attention and understanding, even among the experienced.

What Are We Looking For?

To understand local behavior with respect to food production and consumption, observe these areas:

1. What is the money flow in the food system? What credit system is used? Are farmers truly independent, or are they dependent through debt, or laborers on land owned by others? Is there a social reciprocity system that reinforces a dependence situation? And on whom are they dependent? Can they afford the extra inputs to invest in new seeds, techniques, or equipment that would ultimately recover more grain?

2. What is the belief system of the people regarding food supply? Do they see it as a purely commercial transaction or are supernatural forces involved?

3. Do they understand the connection of more food with better nutrition and health, ie, do they see themselves as having some control over their health?

4. What are the social connections to securing and consuming food? Is much food given away, or eaten in larger gatherings, and how would that affect the costs, risks, and benefits of saving more food? Can social obligations be used by hungry people to buy food, and thus give more incentive to grain recovery? Food has many social and personal functions in addition to nutrition and these should be well understood so that suggested changes permit continuity of these functions.

5. What do the people do with extra money? If saved grain is sold for cash, then the saving may be less critical. If extra money opens problems of taxes or extra grain opens increased obligations within a social reciprocity system, a saving may be disadvantageous to the grain owner.

Other questions and attitudes are explored in Part B of this chapter.

Social and Economic Ecology

Even with current ecological awareness, it may still be necessary to recognize the interrelation that exists. The facts of ecology are well known for animals and plants and the physical environment, but are surprisingly neglected in the social and economic spheres. There are social and economic ecologies, too, and the effects of a survey or proposed change are felt in many ways, and among many people other than those directly involved.

Social ecology may be linked to economics, if economics is broadly defined to include all actions that maximize security and the ability to cope with one's surroundings. People relate to one another, form and break alliances, cooperate and compete. Some hope only to stay alive, to break even with life, while others — more and more as the potential for change becomes known — try to improve their levels of wealth, power, and prestige. The individual entrepreneur, in fact, may well be a role learned from colonials, who brought with them the idea that work and intelligence (cleverness) can raise a person from low to high in a lifetime — a phenomenon previously seen only via miracles and natural events, not under one's own control.

To understand social ecology, it is useful to describe levels of wealth and power within a community and to learn the paths by which people can get there. Some positions will be very stable, others precarious, and the degree of stability should be noted as well. Then, the effects of a study or a proposed change can be cast against this background: What will happen to X if we do this? Or how does X see this change as affecting his community and his position? Remember that he may see the exercise from a different vantage point than that of the investigator.

It also helps to learn how people define security, what their real aims are, and whether they understand that they can better their lot without incurring enemies who now have less. Competition may be based on the philosophy that if I get more, someone else will get less. Riches breed anxiety in such a system, and it serves as a device to inhibit excessive differentials.

Local social customs define associations. Such customs act as social glue to serve as markers of who belongs where, or who wants to move where, or who can trust whom, or what set of rules a person is following. Customs can also define social boundaries to identify different groups within a community.

Economic ecology can also be viewed in numbers. This is the grain pipeline, but determined from farmer to consumer, with attention paid to debts incurred and values received along the line, not only in money but also in services and promises of services. Prices may be less to one person than another; that is not always unfair, as it may be the seller's way of repaying a debt or earning a future favor. Credit is all-important in understanding the pipeline as the farmer's actions may well be linked to his credit sources and their limits.

Another socioeconomic factor is visible difference. A man may *not* want to do better than the others, at least visibly, if envy is to be avoided. In some societies, invisible success is tolerated but in others it is betrayal of the common good, and only a cooperative or communal effort will work, as no one would be obviously climbing over the others. A knowledge of attitudes toward envy and success should be useful in planning the scope of proposed changes.

Outside development processes have reached almost everywhere in the world and the remembered effects of local involvement have not been universally favorable or unfavorable. Onset of a new program, either survey or direct assistance, is an intervention into today and brings with it future concerns. The investigator will get more done more accurately when he knows the actions and interactions of the people he is working with, when he recognizes the similarities and differences among them, and when he knows where they have been and which way they are going.

CHAPTER III

B. Anthropological Signposts

C. C. Reining

The researcher or project manager needs a clear understanding of the cultural and social setting in order to meaningfully assess grain losses. At its most basic level this means knowing who does what to the grain, how, when, and why. It is easy to see that measurements of tangibles should never lose sight of the people who produce, process, and consume those tangibles. However, there is a need for understanding the human social and cultural factors which go far beyond that immediate level and which will dramatically influence the degree of success of a loss assessment effort.

Because so often the project managers in grain loss programs are outsiders to the area being studied, there may be a high incidence of cross-cultural communication gaps which can impair the progress and accuracy of loss surveys. However, with careful effort, much can be done to overcome such cultural perception difficulties. As cross-cultural communication gaps are likely to occur throughout the span of the project, the effort and time spent in developing a cultural understanding will more than repay itself in later-saved time and expense.

Good social and cross-cultural communication skills will be required in selecting, training, and supervising field-workers; in determining what questions need to and can be asked in field surveys, and in ascertaining how to phrase them for ease of comprehension; in identifying which individuals are the best informants for specific questions; and in allowing for and putting into proper perspective potential biases including those of the local farmers, grain handlers, extension workers, field investigators, and the project manager himself. Particular objectivity will be needed when local ideas and values differ from those of the investigator.

The continual need to balance and blend technically ideal procedures and approaches with social, cultural, and political realities is a process which will influence conscious and unconscious cultural values and perceptions. More than any other discipline or subject area involved in grain loss assessment and reduction, the sociocultural one lends itself least well to a step-by-step or procedural treatment in this manual. The cultural observation guides provided at the conclusion of this chapter should not mislead the reader. No such guide could be comprehensive. The guides presented here are provided as a tool — a thought-provoking means of helping project managers and their personnel to formulate their own process for understanding the salient aspects of the local culture and to develop the greatest possible depth of understanding.

In many circumstances, the limited time available for survey planning will make invaluable the short-term services of expert anthropological or sociological assistance. It is assumed that every project would benefit from the assistance of such expert staff members, although the reality of limited project funds and personnel will often mean that such professional assistance will be brief. Where such assistance is not available, a suggested analytical tool for identifying the human element in the grain pipeline is in following through each relevant process or stage in the pipeline to trace what might be called the "grain handlers' pipeline." This can be usefully broken down as to who (age, sex, and social position) does what, when, where, and why. As the situation is studied in more depth, critical and subtle elements will become clear, including who has the decision-making authority and which individuals might be most and least amenable to changes in their present grain handling and storage procedures.

In spite of recent widespread recognition that women's roles in developing countries have been largely overlooked, it is useful to emphasize this issue again here. In subsistence farming cultures, women often perform many of the tasks in grain handling and storage. Too often researchers and project planners have failed to see and describe the role played by women. As a result, vital parts of the intricately interwoven cultural framework have remained unobserved and unaccounted for, only to be unpredictably changed, alienated, or harmed when programs are initiated to improve the situation.

An outsider, defined as any person who does not live in the community, finds it difficult to find out who does what, why, how, and when. When the investigator is a man and the major tasks are performed by women, the problems for an unknowing man can be insurmountable. It is not satisfactory to ask the men of the village what the women do, how they do it, when they do it, and why. It is not uncommon to have the men say that a certain task is done a certain way, and to find out later that their perceptions are off, when the task is performed by women. In addition to men's lack of awareness about particular details of women's work, one must add the cultural constraints imposed on outsiders, particularly those who are men, in communicating directly with the women. This takes time and carefully selected and well-prepared investigators.

Female survey workers may be necessary in some cultures to gain access to women. However, it is overly simplistic to assume that a female worker will necessarily be more perceptive or reliable than a male in specific women-oriented investigational work. If there is a severe access problem in outsider men even being able to talk to women, it may be essential to have female investigators, although in selecting field investigators, the more perceptive, imaginative, reliable worker is always preferable, whether male or female.

When project managers and their field-workers do not speak the same language and especially when there is a marked difference in their cultural or social backgrounds, the inevitable communication problems caused by translation and cultural differences need to be recognized and dealt with. Field-workers' understanding of instructions and the reliability of their observations must be carefully verified. This verification needs to be done in a number of ways:

1. Regular personal observation in the field by project managers to check on workers' methods and reliability.
2. Rephrase questions and instructions to assure full understanding and accurate communication between director and workers.
3. Check several sources of information for cross-checking of observations and assumptions.
4. Get to know field-workers' ways of thinking, biases, weaknesses, etc.
5. Keep to a minimum the number of intermediaries between project direc-

tor and the village situation, to minimize communication problems and distortion of information.

In summary, it would be hard to overestimate the importance of social and cultural awareness and understanding on the part of loss assessment project managers and their personnel. Personal flexibility and willingness to learn will be great assets in order to gain this understanding. Countless decisions will be made which draw on this cultural understanding in balancing and adapting the project's technical needs and scientific ideals with social and cultural realities.

The following cultural observation guides are intended to help bring to light salient cultural factors, although no amount of study and instruction will replace the learning opportunity of direct, personal experience in living and working in a cross-cultural setting.

1. Social Organization
   a. Describe the levels of wealth, power, and prestige in the community. (Comment: Relations between social classes can have a profound effect on handling basic items such as grains.)
   b. Who and what comprises the basic production unit?
   c. Who and what comprises the basic consumption unit?
   d. If they are not the same, why is there a difference?
   e. How do these units form into larger units?
   f. What are the local names of these units and do they have meanings?
   g. Which persons or positions are the leaders within each level and how do they communicate?
   h. Who does the harvesting, transporting, drying and other preparation, and storing?
   i. Who removes grain for sale or consumption?
   j. Who has control of the grain before and after storage?
   k. What is the relation between producers or producing units and purchasers of the grain?
   l. Are there any legal restrictions on the sale or transport of grain?
   m. What are the differences in storage of grains intended for sale as compared to those intended for home consumption and for seed?
   n. If there are crops intended entirely for sale, what are the differences in responsibilities and in handling?
   o. What types of occupational specialists are involved in the grain production and storage?
   p. Who obtains the materials for storage facilities?
   q. Who builds the storage facilities?

2. Domestic Organization
   a. How large is the usual household and what kinds of relatives does it contain?
   b. Does it contain any unrelated persons, such as permanent servants or temporary laborers?
   c. Is the household the basic unit or a subunit of production and/or consumption?
   d. How does the household link with the rest of the community?
   e. What kinds of work are usually done by women?
   f. What kinds of activities are avoided by, or restricted for, women?

    g. What kinds of work are usually done by men?

    h. What kinds of activities are avoided by, or restricted for, men?

    i.   Who makes the decisions about the various stages of production, storage, processing, and sale or consumption of grains in the household?

    j.   Can exceptions be made to the rules about who makes the decisions and under what circumstances?

    k. Who does the training in storage techniques?

    l.   What happens to stored grain in the event of death(s)?

    m. How is transfer of authority made on the death of heads of consuming and/or producing units?

3. Cultural Factors

    a. Are losses permitted because of lack of awareness?

    b. Are losses felt to be inevitable?

    c. Are the people concerned about their grain losses?

    d. What do they think should be done and why haven't they done it?

    e. Which grains do the people believe store the best or longest?

    f.   Which grains do they believe are hard to store?

    g. How do they explain the differences in storage characteristics?

    h. How do they accommodate these differences? Do they have different methods? Do they consume some grains more quickly than others?

    i.   How does the availability of other crops, such as root crops, influence the storage of grains?

    j.   What are the indigenous materials used to help prevent damage to stored grain?

    k. What do the people see as the tangible causes of damage to stored grain?

    l.   What are felt to be the intangible or supernatural forces controlling losses?

    m. How do they attempt to influence both the tangible and intangible factors?

        (Comment: There are serious problems of categorization here, both in Western and indigenous terms. Often the distinction between "magical" and "scientific" becomes blurred, as when a local remedy that is felt to have mostly spiritual qualities may, in fact, have demonstrable effects on stored grain, while other devices believed to have more direct effects do not have any discernible ones. Most preventative practices are a blend of empiricism and mysticism.)

    n. What will be eaten that might have been damaged?

    o. What are the local guidelines for what should and should not be eaten?

    p. What is done with spoiled grain? For example, is it fed to chickens or other domestic animals?

4. Transition and Change

    a. Is a need for change or improvement felt by the local people?

    b. How do they want to change the situation?

    c. Is their knowledge of desired change sound enough to understand the ramifications?

    d. Can they afford the new materials?

    e. Will they be able to sustain the new equipment and techniques?

f. How do innovations get into the community? Are there key positions or individuals for introducing innovations?

g. What improved procedures have been introduced? By whom? Successfully?

h. Have storage systems of various indigenous systems in the *same kind* of environment been compared?

(Comment: Most communities have had a long time to experiment with adapting to their particular setting. It is usually difficult to improve upon the local arrangements given the resources available. If introduction of new techniques is deemed necessary, it may be more effective to consider transfer from a similar indigenous setting rather than from Western culture.)

5. Individual Factors

a. The local person

i. How typical is the person supplying the information?

(Comment: Often the typical or normal person is too busy to want to spend time talking with outsiders. The persons most available too often are marginal to the community.)

ii. What does the informant see himself or herself getting from the interview?

(Comment: It is very human to constantly assess any situation to maximize the returns. Beware of creating false hopes.)

iii. What are the biases and interests of the interviewee?

iv. Is the interviewee skewing the information to fit the situation as perceived?

(Comment: There is often a tendency to tell the interviewer what the interviewee thinks he wants to hear. Misunderstanding is altogether too frequent. Consider the difference in response if the interviewee thinks there may be a tax imposed on the stored grain, as compared to the impression that compensation may be paid for lost grain.)

v. Are the interviewees saying what should be rather than what is actually the case?

(Comment: It is important to distinguish between the real and the ideal. Observe what they *do* as well as recording what they *say*.)

b. The interviewer

i. What are the biases of the interviewer?

ii. What are the biases and interests of interpreters, if used?

iii. Are problems perceived from the viewpoint of the interviewer or from that of the interviewee?

IV. REPRESENTATIVE SAMPLING, INTERPRETATION OF RESULTS, ACCURACY, AND RELIABILITY

B. A. Drew, with T. A. Granovsky and C. Lindblad

A. Introduction

Basic Assumptions

Every scientific measurement is based on some kind of assumption regarding the real world about which the measurement is supposed to supply some information. Conducting a survey to measure average grain losses is such a measurement and it is based on the following assumptions:

1. Cultural and economic conditions, level of knowledge of farmers, farming practices, varieties grown, and harvesting and storing practices are essentially uniform throughout the area to be surveyed. If this assumption is to be verified by local observation, one will have to understand the cultural milieu. If it is nonuniform in ways that can possibly affect what is to be studied, sampling becomes more complicated and the advice of experts should be sought.
2. All grain to be considered is stored in the same manner in units of approximately the same size. That is, the largest unit is no larger than five times the smallest. If the size variation is greater, then they should be sampled and analyzed separately as two or more populations.
3. Size of farms is uniform to within a factor of 5. That is, the largest farm is no larger than five times the smallest farm (in area producing crops for storage). Again, if the size variation is greater, then they should be sampled and analyzed separately as two or more populations.

These assumptions limit the survey described to a single stratum. This is all that can be done using the simple sampling plans outlined here. More complicated plans should involve the help of experts in sampling as well as in grain loss assessment.

Uses of Survey Data

In designing a sample plan it is essential to know the purpose or purposes for which the results are to be used. For example, one might wish to determine the calorie losses which are incurred due to parasites, in order to determine whether to supplement the farmers' diet, or one might wish to determine the extent of losses in grain held in storage in order to decide whether to treat it with pesticide. In one case, medical-nutrition concepts are involved; in the other, grain losses.

The ultimate use of the results will influence not only the precision and accuracy which are required, but also what is measured and what additional data must be collected. Thus, the measurements which are made and the ultimate use of the results, including the level of loss that is acceptable, must be decided before the survey is designed.

Editors' note: Given the present refinement of loss assessment methods, it is generally accepted that ± 5% accuracy[2] is the best practical limit which can be expected (with rational allocation of resources and time against the potential value of the reduced grain losses). At the same time, where losses are expected to be 15% or less, a ± 10% accuracy level could all but obscure any meaningful information. *Where such is expected to be the case, rapid expert assessment of critical loss points may be economically justified while an extensive in-depth loss survey is not.* For certain economic evaluations, no less than ± 5% accuracy can be tolerated for analysis to be meaningful.

Determining Area to be Surveyed

In making a survey over a large area such as a whole country or region, the sample population should be divided into parts to reduce the problem to manageable proportions or to obtain a uniform population. This is called multi-stage (stratified) sampling.

In such a situation there are two valid alternatives for sampling a population. These are: To include in the sample of a population all of its subdivisions, or to include a random sample of subdivisions of the population.

Section B presents these sampling methods in detail. The rule for this choice is to take all the subdivisions when there are only a few, say 10 or less. If there are more than 10 subdivisions, then as many as are consistent with available resources should be chosen using random numbers. At such a point knowledge about the differences between particular subdivisions may make a valuable contribution to deciding whether to choose all or a sample of subdivisions. Advice from knowledgable people in this area should be sought.

Types of subdivision are extremely dependent upon the local situation but a country (nation) may be divided on political boundaries such as states or into units based on geographic considerations such as lowlands, uplands, river valleys, and arid regions. The last division would be preferred when knowledge or advice is available about the impact of such conditions upon storage losses. In such a case, resources might be allocated to the various regions in proportion to the likelihood of postharvest losses.

The next subdivision might be on the basis of villages or small administrative or political units. Here the units of the subdivision should be listed and random numbers used to choose as many units as can be measured with available resources. Remember that excessive variations in size of storage unit may require separate analysis of samples as two or more populations.

---

[2]In this manual accuracy is expressed in absolute terms. Thus 20 ± 5% means from 15 to 25%.

If there are different types of stores within the unit (administrative or political unit), then each type of store should be considered as a unit in the next subdivision. It is the last possible subdivision to which this manual refers.

Accuracy

Accuracy of an assessment of grain losses depends on obtaining a truly representative sample and making an accurate measurement on the sample. No matter how accurately one measures a sample in the laboratory, the result will be of little value if the sample is not representative. It is equally pertinent that no matter how representative the sample may be, the final result will reflect all the shortcomings of the laboratory measurement.

CHAPTER IV

B. Probability Samples

Bias

The rest of this section will be devoted to methods to ensure a representative sample and to avoid all sources of systematic error often called bias. If we always sample the best-looking stack in the field, or the one nearest the house, or the one the farmer chooses; if we always take samples right by the entrance into a granary, or where the grain looks good, then we may be putting a bias into the sample. Even if we try to choose in a way to avoid bias we may over-correct. If we try to avoid choosing units that are easy to reach, we may unconsciously choose units that are hard to reach. The only way to avoid bias is to take the choice out of our hands, to give it to a table of random numbers. The method is called "probability sampling," and its result is a "probability sample."

A Random Sample or a Representative Sample?

When establishing a sampling pattern, confusion exists between the terms "representative sample" and "random sample." Representative sample usually refers to a "stratified random sample," in which strata are defined and represented in the sample in proportion to their size in the sampled material.

If 1) the strata have something to do with the property to be measured and if 2) a random sample is taken within each stratum, the variance of the estimate may be lower than that of a completely random sample. Both conditions are necessary, however. The following examples will clarify what is meant by such terms as randomization, stratification, random sample, and stratified random sample.

Randomization or Unrestricted Random Type

25 units

for

potential

sampling

1	2	3	4	5
6	7	8	9	10
11	12	13	14	15
16	17	18	19	20
21	22	23	24	25

1. The area or volume is divided into equal-sized units.
2. The units are numbered consecutively.
3. Units are selected for sampling based on a table of random numbers (Appendix B).

The advantage of this sytem is that each unit has an equal chance of being selected.

Some problems to be encountered are that establishing and setting it up may be difficult; units sampled may be grouped, by chance, in one area or another of the area sampled; and it may get a poor estimate of variance due in part to clumping of sample sites.

Stratified Random Sample

Strata	I	II	III	IV	V
	1	1	1	1	1
	2	2	2	2	2
	3	3	3	3	3
	4	4	4	4	4
	5	5	5	5	5

5 strata of 5 units each for a total of 25 units for potential sampling

1. The area or volume is divided into equal-sized larger units, each of which contains an equal number of smaller units.
2. The units are numbered equally in each stratum.
3. A unit for sampling from each stratum is selected by a table of random numbers (Appendix B).

Some advantages of this system are that locating sampling units is easier, it forces samples not to be grouped, and it gives a better estimate of variance since less clumping occurs.

The two sampling patterns given below are not recommended for use in a loss assessment survey, but are presented for clarity.

Systematic Sample

A sample is taken every so many units, eg, every 10th bag as it is moved from location to location.

Some problems to be encountered are assumed damage or loss is uniformly "normally" distributed, which is rarely true for insect populations, the sampling pattern may conform to some inherent distribution pattern of the damage, and no random component is included and therefore statistical procedures cannot be used.

Centric Systematic Pattern

A sample is taken from the exact center of each unit. If such samples are analyzed using parametric statistics and compared to samples obtained by the random pattern, results may truly reflect what is present.

All problems present with systematic samples are also present with centric systematic pattern.

The sampling patterns illustrate the advantages of having some knowledge about the material to be sampled, and show one way to use such knowledge. But when there is no knowledge from which strata may be deduced, complete randomization is the only way to obtain a representative sample. This applies to each cell or stratum in any scheme of stratification. A random sample should be taken within each cell or stratum. Otherwise, the advantages of stratification may be lost.

Properties of Probability Samples

This section presumes that a probability sampling plan will be used. The reasons for this are:

1. With this type of sample one may calculate confidence limits within which the actual value of the result is reasonably certain to lie.
2. Generally one may determine in advance how many samples must be taken.
3. This type of sample is guaranteed to be representative.

The actual value is the value which would be obtained if the loss *in every unit* in the area were to be determined.

Observational Units

The observational unit is the container, location, or process from which a sample will be removed to determine the loss evident in the sample. This is the smallest division or unit in which grain is held. It might be stacks in a field, small silos or granaries on a farm, or woven baskets. It would be a single basket rather than all of a farmer's storage baskets; it would be individual bags rather than the whole warehouse. Accuracy of the entire survey will depend on the accuracy with which the loss is determined on each observational unit.

To facilitate sampling, the observational unit should be as small as possible. This makes it easier to get a representative sample since it will be possible to mix all the grain thoroughly and reduce the sample taken by quartering or using a sample divider. This may be feasible where the grain is in baskets or in stacks in the field. In silos or granaries it may not be possible and, unless the sampling is done with skill, the sample may contain a systematic error which cannot be removed by any later calculation or analysis.

When any container is sampled as a unit, the assumption is that the defect, contamination, or other characteristic to be determined is uniformly or at least randomly distributed within the unit. As a practical matter such is usually not the case.

Insects/mites, moldy kernels, rodent depredation, and insect-eaten kernels are more usually in location-oriented pockets (see Appendix A).

With time and money constraints and often with cultural-traditional limits also imposed, the best that can be done is to design the mechanical sampling so that the sampled grain will be as representative as practical of both the undamaged material and the layered or pocketed defects.

In any study the investigator needs to report what was done and why so that the significance of the data can be understood by those who will use it.

Where grain is stored in storage units of variable sizes or types, a person with competence in statistics should be called upon to help design the sampling plan.

Number of Samples

To decide roughly how many samples must be taken, two items of information are needed: the desired confidence limits, ie, the estimate of the overall average loss within 1, 2, 5, or 10%, and the range of losses to be expected. The range is the difference (in percent) between the highest expected result and the lowest expected result.

With these two items, one can find from Table II how many observational units will be sampled and measured to get a representative sample. If the number to be sampled is too costly for available resources, the desired confidence limits will have to be lowered. If the range is underestimated, the number of samples taken will be insufficient. Therefore, it is generally recommended to make liberal estimations of the range expected unless the population is well known.

For example, as shown in Table II, if the lowest result that is expected is 25% loss and the highest expected result is 85% loss, then the range is 85 − 25 = 60, and if the desired precision is ± 5% the sample must include at least 81 units. If a sample of 81 units gives a result of 40% loss, the results should be interpreted as 35-45% loss (40 ± 5%).

The above procedure is calculated on American Society of Testing Methods (ASTM) Recommended Practice E122-58 and is based on statistical theory. Other procedures for determining sample numbers which are based on intuition such as arbitrary numbers and square root samples do not allow specifications of desired precision in advance.

Table II is mathematically calculated to assure representative sampling *regardless of total population size*. It is based on the range of results expected and desired confidence limits.

If the actual number of units is less than the number given in the table, then *all* of the units should be sampled.

Preliminary Surveys

A preliminary rapid fact-finding survey, mentioned in several places in this manual, is of value in gathering information to assess the homogeneity-nonhomogeneity of the system.

Answers to the following kinds of questions should be obtained by the preliminary survey:

- Are there large differences in culture? Income level? Farming, harvesting, drying, storage practices? Crop and variety grown?
- In what size unit is grain stored? What is the largest unit found? The smallest? How many of each class?

TABLE II

Required Number of Samples

		Range of Results Expected							
		100 (%)	80 (%)	60 (%)	50 (%)	40 (%)	30 (%)	20 (%)	10 (%)	5 (%)
	± 1%	5,625	3,600	2,025		900		225		
Desired	± 2%	1,406	900	507		225		57		
Precision	± 5%	225	144	81		36		9		. . .
	± 10%	57	36	21		9		3	. . .	. . .

Note: This table was derived by standard calculations based on a conservative estimate of population-defined standard deviation = range/4.

Sample numbers in this table were calculated using eq. 1 in Recommended practice for choice of sample size to estimate the average quality of a lot or process, ASTM E122-58, American Society for Testing Materials (1958).

- How big is the largest farm (village)? The smallest? How much land does each actually cultivate with crops that will be stored? Can you make a list of all the farms? Can you locate them on a map?
- How many storage units of each size class are there on the biggest farm? On the smallest? Can you estimate the number on an average farm?

It may be of value to collect other data in a preliminary survey to facilitate subdivisions into strata or for other purposes. As the preliminary survey uncovers separate strata, it uncovers material that needs to be sampled separately if adequate overall coverage is to be obtained. It is also necessary to look at the total situation (eg, the subsistence or the marketing systems) and then determine what elements are to be measured. In other words, what components do matter? What are the expected ranges of the variables? What should be ignored as trivial?

One needs to know all of the possible ways the population stratifies: geographically, climatologically, politically, and culturally (size of installation, wealth, mechanization, kinds of storage).

The pipeline concept (see Chapter II) is a means of sorting out, for example, situations, locations, economic and political factors. It is a means of focusing on a situation to reduce the study to a homogeneous stratum.

Designing the Probability Sample

To design a probability sample, it is necessary to use a method that ensures that every observational unit in the area to be surveyed has a known probability to be included. When it is known in advance how many units there are and where each one is, then a list is made and the units are each given a number in series from one on up to the total number. Then a table of random numbers (see Appendix B) is used and those locations whose numbers come up are sampled and measured until the required number have been done.

If the number of units and their locations are not known, an estimate of the total number of units from the preliminary survey can be used to calculate what proportion of all units to sample. For example, if one wants to sample 200 units and he estimates that the area to be sampled may contain 2,000 units, then he takes one unit chosen at random for every ten units found. A method for doing this is to make up lists of random numbers for farms containing various numbers of units and put them in envelopes for the sampler. When he comes to a farm that has 51 units, he first numbers each of the 51 units. Then he opens an envelope labeled "45 to 51" which contains five random numbers (between 1 and 51 inclusive). He then takes samples from the five units given.

In sampling farms if the number and location of farms are known, each farm is given a number and the farms to be visited are chosen with the table of random numbers.

Taking samples on a farm which has more than one stack or granary should also be done at random, taking into account any known pattern of use or any other known nonhomogeneity. It is best to decide in advance how many units will be sampled on a farm and to have sets of random numbers of the correct size in envelopes. Then the sampler can number the units (baskets, stacks) found, and choose an envelope labeled for that many units that contains the required random numbers (see Appendix B).

Note: In sampling it is always a good precaution to identify extra sampling points and to take samples from these sites to replace the inevitable accidents, dropouts, or loss of sampling sites.

CHAPTER IV

C. Detailed Instructions

Choosing Farms or Villages

All the farms (villages) in the area to be surveyed should be listed and the number of samples that are required should be determined (see Table II).

If there are more farms than samples required, and if the farms are all the same size (within a factor of 5), then

- Give each farm a number from 1 to as high as necessary.
- Use a table of random numbers to choose the farms to be sampled. The farms chosen may be visited in any order that is convenient.
- Obtain samples from one observational unit (stack, basket, crib, etc.) on each farm. Choose the unit with random numbers after seeing how many units there are on the farm.

If more samples are required than there are farms, and if the farms are all the same size (within a factor of 5), then

- Determine (or estimate) how many observational units there are in the area to be surveyed. The total number of units is called N and will be greater than the number of farms, if several observational units are present on each farm.
- Determine the number of samples necessary from Table II. This is n. The fraction n/N is the sampling proportion.
- On each farm (or in each village) count the number of observational units and multiply by the sampling proportion. The result, rounded to the next highest whole number, is the number of units to be sampled.

Sampling on Farm or in Villages

Labeling of Samples. All samples must be labeled and retain their identity as to date collected, exact location of source, how sample was obtained, grain type, variety (if known), time in storage, and type of storage.

Procedures for Sampling

Standing Grain in the Field

- Choose an area (in square meters in broadcast crops or linear area in row crops) that will yield 1 to 1.5 kg of shelled grain.
- Divide the field into units of the chosen area.
- Give each unit a number starting with 1 and going as high as necessary.
- Choose as many random numbers from the table furnished as there are samples to be taken.
- Harvest and shell the grain in the unit areas whose numbers were chosen.
- Package the grain from each unit for transmission to the laboratory.

In the Field in Stacks (If Each Stack Contains More Than 2 kg of Shelled Grain)

- Give each stack a number starting with 1 and going as high as necessary.
- Choose as many random numbers from the table furnished as there are samples to be taken.

- Shell each stack whose number was chosen.
- Reduce the grain by coning and quartering or by using a sample divider (see Appendix A) to a sample of 1.5 kg.
- Package the sample for transmission to the laboratory.

Note: If each stack contains less than 2 kg of shelled grain, choose twice as many random numbers as there are samples to be taken. Combine the grain from two stacks into a single sample for transmission to the laboratory.

When the Shelled Grain is Stored in Baskets
- Give each basket a number starting with 1 and going as high as necessary.
- Choose as many random numbers as there are samples to be taken.
- Reduce by coning and quartering (or use a sample divider) each basket whose number is drawn to a sample of 1 to 1.5 kg.
- Package the sample from each basket for transmission to the laboratory.

When the Unshelled Grain is Stored in Small Units (Such as Baskets and Bags). If the grain is stored in small units on the cob, head, or panicle, shell the contents of the whole unit before coning and quartering to yield a 1- to 1.5-kg sample.

When the Unshelled Grain is Stored in Large Cribs, Silos, or Granaries. To sample grain stored unshelled in cribs, silos, or granaries, unload and shell the entire lot. Then cone and quarter (or use a sample divider) to obtain a sample of 1 to 1.5 kg. Or unload the grain equally into baskets and then use the method for unshelled small units (choosing baskets by stratified random sample).

Note: In storage, ears of cob maize or panicles of sorghum/millet and maize can be labeled randomly as the crib is filled. The farmer can then be asked to set these ears aside as he encounters them during emptying. Determining an adequate sample of ears or heads from a crib can be a problem, however. This procedure should be used only after careful study of its applicability to the local situation.

Large Bulk Storage Units, Shelled. Obtaining a representative sample from a large bulk container is difficult. Ideally the grain would be transferred into another container in such a way that samples could be obtained from the grain as it falls into the new container. A container small enough to be handled easily should catch the entire falling grain stream until it is full or passed through the entire stream and the caught grain placed into a larger sample container. This procedure would be repeated at frequent, regular times throughout the transfer.

When all the grain has been transferred, the sample that has been collected may be reduced by coning and quartering or by using a sample divider to 1 to 1.5 kg for transmission to the laboratory.

If it is not possible to sample the grain during a transfer, then a probe may be used. It is recognized from research results that a probe sample is not representative (see Appendix A). When probe sampling is used a note should be made of that fact in the final report. In using the probe, an effort should be made to reach every part of the storage container. Several times as much grain as is necessary for the final sample should be taken and then reduced by coning and quartering or by using a sample divider. Samples should be taken with the probe *in at least* the positions shown in Fig. 6, using a compartmented probe that samples at all levels.

Mass Storage in Bags. Obtaining a representative sample of a large mass of grain stored in bags can only be done if every bag is accessible. To sample such a store requires that one chose enough random numbers and then move the grain one bag at a time to a new location diverting bags for sampling corresponding to the random numbers. The diverted bags should be sampled, preferably by coning and quartering the whole bag or putting it through a sample divider to obtain 1 to 1.5 kg of sample for the laboratory. The remainder can be returned to the bag and to the store.

A less satisfactory alternative is to obtain a sample from each randomly chosen bag by probing. A probe long enough to reach diagonally from corner to corner of the bag should be used and the bag should be probed on both diagonals and in enough other locations to obtain 1 to 1.5 kg of grain from each bag.

It should be noted if every bag is not available to be sampled so the result will refer only to those bags that were accessible. The bags sampled should be chosen by assigning numbers to those that are available and using a table of random numbers to choose the bags.

Sampling procedures should always be reported, especially when the sampling is suspected to be nonrepresentative as in the case of stacked bags, unshucked or unshelled grain heads and cobs, and when there are visually observed concentrations of insects or mold, or both.

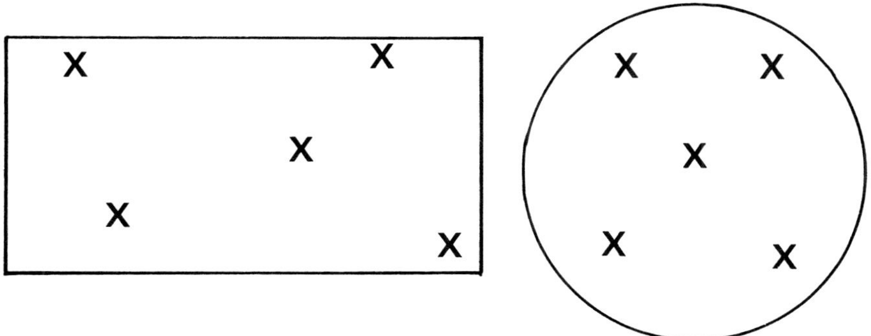

Fig. 6. Probing locations in rectangular and round bins.

V. LOSS MEASUREMENTS AS RELATED TO SITUATIONS WHERE THEY OCCUR

Many, if not most, postharvest losses occur as a result of externally applied adverse factors, as when insects, rodents, and birds consume the grain. Other losses occur while, or because, the grain is in an otherwise useful state or process. Losses are often sustained while the grain is being threshed. These losses are brought about by (deficiencies in) the threshing process.

Grain must be transported from farm to urban centers. During this process, bags or vehicles may leak and grain is lost along the way. The transporting process is useful; it also may result in losses.

In this section, measurement procedures are dealt with as they relate to the process the grain is undergoing. The techniques for analytical-type testing not given herein are in Chapter VI.

Processing losses are affected by prior induced quality factors such as checking and cracking rice and corn, and a methodology should put such factors in perspective.

Methods are not given for all the procedures needed to determine prior-to-processing damage that brings about subsequent losses during processing. Also methods are not given for all processing damage that causes losses during further manipulation.

A. Background Information

D.A.V. Dendy, with K. L. Harris

Two basic concepts are used in this chapter. One is to measure the situation (usually output) of a given operation and to compare it with an ideal (hand or special machine) operation. The other is to measure losses by weighing the various food, feed, and other streams and making direct calculations of what does not end up as food.

Whether the loss is waste is not a matter that depends on methodology. Bran can be waste, feed, or food, independent of loss-assessment methodology.

What results as food may be compared to total food value, to food obtained by the best possible process or best possible commercial process, or even by an experimental process. The methodology simply needs to be set up to make the required measurements.

Shelling of Maize

Stripping of maize grain from the cob is known as shelling. Losses occur wherever mechanical shelling is not followed by hand-stripping of the grains remaining on the cob. Certain shellers damage the grain, making insect penetration easier and subsequent storage losses higher.

Threshing

Losses occur during threshing by spillage, by incomplete removal of grain from stalk, or by damage to grain during threshing. They also occur after threshing due to poor separation of grain during cleaning or winnowing.

Incomplete stripping usually occurs in regions of relatively high labor cost at harvest time, where the method of threshing leaves some grain unthreshed but labor is too expensive to justify hand-stripping. Workers in Malaysia observed that 1.13% of paddy was lost by falling outside the threshing tub; it was also noted that up to 11.7% was left on the straw.

Certain mechanical threshers have cleaning equipment designed for only dry grain. A wet season's harvest, eg, of paddy, will clog the screens and grain will be lost with leaf and broken stalk.

Use of oxen for threshing paddy provides a trodden straw said to be more easily digested. If the threshing floor is muddy or cracked, grain will be lost.

There can be a 5% increase in cracked and broken kernels after combine-harvesting paddy compared to hand-harvesting and hand-stripping.

Cleaning and Winnowing

Cleaning is customary before milling. At the home, hand-cleaning is a combination of hand winnowing with hand removal (eg, of stones); losses can be very low when carefully done or high when siftings are allowed to scatter on the ground or winnowing done with the same result. With correct equipment, losses should be low in mills, but equipment undersized for the quantity of extraneous material, such as dirt, will cause losses of grain by removal with the dirt or by the dirt being carried forward into the milling stages. Loss assessment is difficult as losses are usually low; high losses are spotted by operators and the extraneous matter is recleaned.

Drying

Two losses are frequently caused by drying: removal of grain and portions of grain from the drying system, and damage to the grain leading to a subsequent loss.

Grain which is dried in yards, on warehouse floors, or on roads will be partially consumed by birds and rodents. Wind, either natural or from passing vehicles in the case of road drying, will blow some grain away. Although very little grain is removed on vehicle tires, damage by vehicles may cause subsequent losses. Mechanical dryers may cause damage leading to removal of parts of the grain (such as bran) from the system either in the air flow or in subsequent cleaning operations.

The principal loss-factor occurring during drying is caused by kernel cracking ("checking") of grains such as rice, which are eaten whole. Usually the greatest damage occurs through re-wetting which happens when grains of different moisture content are mixed in a dryer, and when rain or dew re-wets grain in a yard. The damage is manifested as broken grains during milling, especially in the polishers.

Primary Processing (Milling)

This includes all processing operations carried out on grain in the home or mill, such as cleaning, parboiling, hulling, de-branning, grinding, and separating (classifying). Secondary processing (cooking, baking, fermenting, extruding) is excluded; such losses as occur are usually unavoidable, being intrinsic to the process and preventable only by a change of process — more a subject for the sociologist than technologist.

In the home and small mill, grain processing is effectively a batch process in which relatively small quantities of grain are processed by one or more operations and the product collated, then brought together for sale or other processing. In large mills, the processes are continuous and loss measurement is performed periodically by sampling product streams. All of the pre-milling history affects the fate of the grain during milling.

Parboiling

Though easily quantifiable losses of soluble materials occur during parboiling of paddy, these losses are more than offset by the improvement in nutritional value of the kernel.

Hulling, Polishing, Especially Rice Milling

Removal of the outer coats from a grain may take place in one or more stages. For paddy rice, red sorghum, and oats, considerable mechanical effort is needed to remove these layers. Any weakness in the kernel, caused previously or inherent, will manifest itself at this stage. Even with grain in perfect condition, only the best process with correctly set machinery will yield an out-turn of whole polished grains approaching 100%. In the case of rice, broken grains command lower prices and finely shattered material ceases to be human food. Some leaves the mill in the husk (fuel or waste), but most with the bran (feed). Bran removal may be considered a loss. With the consumer demanding rice with a high degree of polish, the loss at that stage must be measured and then changes made to keep the losses to a minimum. It has been noted that even a 1% increase in yield of whole grain rice can result in huge increases in national resources.

Grinding

In some processes such as wheat milling, removal of an edible part of the grain, eg, the germ, is deliberate and desired by the consumer. Whether this is a loss depends on the terms of any particular study. However, mechanical losses of desired ground products frequently occur, often caused by maloperation of the process or worn equipment. Common processes are pounding in a mortar, grinding between stones or toothed steel plates, and the complex Hungarian system for milling wheat into flour.

Separation

Whether the separation of edible from less desired products is done in the home (eg, winnowing hulls and bran from rice) or mill (eg, sieving flour from bran), complete separation is rarely achieved. With rice, it is difficult to separate the more finely broken grains from bran, and with wheat, flour adheres to bran and special equipment is used to remove most of this as flour.

Nonuniformity

Processing of mixtures that are nonuniform because of such factors as hardness and softness of kernels, size (length, plumpness, etc.), and moisture content difference is itself a cause of losses.

CHAPTER V

B. Guidelines for Performing Studies of Farm Storage Losses[3]

J. M. Adams and G. W. Harman

1. An inter-disciplinary team, comprising at least a storage technologist and an economist, is necessary. The team should arrive in the area early enough before harvest to enable it to plan effectively, to select fieldwork areas, to train and brief enumerators, and to conduct necessary trial runs.

2. The sampling frame for investigations on both technical and economic aspects should be determined and stratified. Areas chosen for fieldwork should be as representative as possible of traditional practices, both preharvest and particularly postharvest. (See Chapter IV.)

Information on the technical aspects of losses should be obtained by:

1. Collecting the necessary baseline data on the moisture content, damage, and bulk density (bushel weight) of the commodity immediately prior to storage, and recording any procedures involving selection or treatment of the product for storage.

2. Recording the quantity of the commodity placed in storage.

3. Recording the date on which some of the commodity is first removed from the store. Thereafter samples of the commodity should be taken at monthly intervals. The sampling method used should be pre-tested, prior to large-scale use, for its acceptability to both the investigator and the farmer.

4. Collecting information on the rate of consumption of the stored commodity over the storage period. This should be done on each sampling visit.

5. Analyzing the samples to obtain estimates of loss and applying these to the consumption pattern to obtain an estimate of loss over the complete storage period. Weight of a standard volume of grain corrected for moisture content changes should be used to assess losses in samples when regular sampling is performed. If this is not possible the formula method may be used to estimate losses within individual samples, but with less accuracy. (See Chapter VI.)

6. Setting up simulation stores, if necessary, which are under the control of the investigator and simulate the farmers' pattern of consumption. The commodity should be accurately weighed in and out of the store. Care should be taken that the grain placed in these stores is of the same quality and selected in the same way as that placed in the farmers' stores.

Information on economic aspects will be obtained:

1. By a questionnaire survey on a once-only basis, conducted with a representative sample of farmers.

2. On a regular basis from farmers from whom grain samples are taken, if this is part of the research, and from official sources.

The questionnaire survey should be evolved in three stages:

1. A basic outline following on-site discussions.

[3]Adapted from J. M. Adams and G. W. Harman. The evaluation of losses in maize stored on a selection of small farms in Zambia with particular reference to the development of methodology. Trop. Prod. Inst. Rep. G109 (1977).

2. A trial run (see below).

3. A final revision. The questions to be asked will depend on the objective of the survey, the potential ability of the interviewees to respond, and the time and staff resources available to the research team.

The questionnaire should be sectionalized as required by the study. The following is a guide to some but not all of the main subject areas:

- General. Farmer's status, household size, measurements of wealth (cattle ownership, alternative employment, size of farm), credit facilities and usage of.
- Cropping. Crops grown, area, and disposal/storage.
- Principal grain crop(s) production. Varieties grown, seed source and costs, use of fertilizers and insecticides, drying and pre-storage activities.
- Storage. Quantity stored, form in which stored, number and type and structure of stores, cost of stores and store materials, labor for building and maintenance, age of stores, potential life, pre-storage and in-store treatments, dates of first and last removals, frequency and quality of grain removed, site of removal from the store, usage of grain removed.
- Storage losses. Cause, severity, usage of damaged grain.
- Marketing. Sales of grain which is never stored, quantity, variety sold, reasons for sales, grade/price made, buyers, transportation.
- Buying. Quantities bought, form (grain, meal, etc.), frequency, price, source, usage.

It is important to emphasize that the above are broad outlines only. Each situation may require some addition or deletion and all situations will require precise framing of the questions to be asked. These six criteria should be observed:

1. do not ask unnecessary questions; limit the number and complexity of questions so that each interview is completed in 30 to 40 minutes maximum.

2. as far as possible, frame the questions so that the answer is yes or no.

3. have a trial run and revise or eliminate difficult questions.

4. avoid sensitive questions if possible and seek local advice as to which questions are sensitive. (It is, however, surprising how many seemingly sensitive questions can be asked and will be answered if correctly phrased and properly put, emphasizing the importance of enumerator training.)

5. train enumerators thoroughly, work with them through their initial field operations, and spot-check their activities at intervals.

6. consider the feasibility and advisability of moving enumerators between areas and strata both as a check and as a stimulus on the individuals' performance.

This questionnaire survey will probably be asked of a larger sample of farmers than the one from which samples of the grain are drawn for analytical purposes (assuming that the latter is part of the study involved). Nevertheless, all of the latter should be asked the questionnaire survey; their actual activities on grain removal can be observed in practice and comparisons of observations and statements will provide a valuable check on farmers who are involved in making statements in the questionnaire survey only.

Economic information should be collected on a continuing basis from farmers. If, as is likely, it is necessary to undertake a program of regular sampling of farmers' stored grain, regular visits should be made to collect economic information of usage patterns, quantities and prices for sales and purchases, time required for store building and maintenance work, and cost of materials used.

CHAPTER V

C. Procedures for Measuring Losses Occurring During or Caused by Processing Including Threshing, Drying, and Milling of Most Grains, but not Maize or Pulses/Groundnuts

D.A.V. Dendy, with K. L. Harris

Processes may be continuous or batch. In the former, samples of input and output should be taken at regular and measured intervals. The amount (1, 5, or 10 min) of production taken from various lines in the system can be weighed to give the quantity of stock carried in that line in proportion to other lines. Samples may be taken in the usual way from the bags of grain entering the process and bags of product(s) leaving. Overall mass balances must be measured and converted to standard moisture content or to dry weight.

Two fundamental methods are used: measurement of total system (mass balance), and comparison with a standard.

Measurement of total system. The loss itself may be weighed. The optimum process gives zero loss. Examples are threshing (loss on stalk) and maize shelling (loss on cob). In some cases the loss itself cannot be measured, but the input of grain and output of products can be weighed, the difference being the loss. In other cases, loss will be a comparison of the traditional or commercial system as against a perfect hand-stripping standard.

Comparison with laboratory standard. Comparison is not against a perfect (100% recovery) standard but with an optimum standard, usually taking each unit operation (stage) separately. Although this method is not ideal, if the standard of comparison is adequately described, the comparison will produce useful information.

It is important also that unit operations (eg, hulling and polishing) subsequent to that under consideration (eg, drying) be investigated or that information be obtained on the entire flow in the best possible and most standardized way.

Sampling (see also Chapter III)

Sampling procedures are simple for batch processes such as are carried out in small mills and homes. If a loss of material is looked for, then a weigh-in weigh-out procedure will be adopted. Where a lowering of quality is suspected, a sample should be taken before the process and put through a parallel but optimum process (eg, in a laboratory mill) to compare the products. In continuous systems, the unit operation (stage) can be scrutinized while representative samples of substrate are taken at regular intervals before and after. The condition of the inputs and outputs is determined by laboratory examination. The amount (weight) of the outputs is obtained by comparing the total weight of the streams over a fixed period of time so that the comparative amounts of grain going to food, feed, waste, etc., can be determined. For example, in a continuous flour-milling operation, weights taken over a 1-min period of flour, bran, shorts, and dust will show what proportion goes into each product. If dust is 0.5% of the flour + bran + shorts, and dust is used for fuel while flour, bran, and shorts are all food, then the loss in this stage is 0.5%.

Operators

Where losses depend on operator efficiency, there will always be the problem of deciding whether the operator is working normally or at an enhanced efficiency to impress the assessor. The tester must gain the operator's confidence and impress on him that it is not he who is under scrutiny.

The following examples can be used as a guide for other unit operations.

THRESHING LOSS 1: Unstripped Grain (Loss With the Straw)

A suggested method is as follows. Random samples of bundles of cut grain are chosen and threshed by the customary method. The threshed grain (sample 1) and straw are retained. Directly supervised labor hand-strips every grain (sample 2) from and out of the straw. The two grain samples are then hand-winnowed carefully to bring hand-stripped and mechanical material to the same quality. The good grain is weighed, moisture content measured, and the weights converted to a standard moisture content.

It is important to examine the two samples and estimate as accurately as possible (eg, by hand sorting of a representative subsample) the proportion of useful quality grain. Note and record unfilled, immature, or green grains that would be rejected during subsequent processing. Then the total of these plus extraneous matter should be determined and the estimated total weight subtracted respectively from the main threshed sample and the hand-stripped material. The good hand-stripped grain would normally be lost, and the loss is the percentage ratio of this to the total good grain, hand-stripped plus normally threshed.

Losses due to scattering and spillage, which may occur with certain threshing procedures, would be evaluated separately by recovering scattered or spilled grain from known or controlled amounts of threshed grain or by weigh-ins and weigh-outs if these are known or can be determined.

THRESHING LOSS 2: Damage to Grain

The method to be followed for estimating grain damage during threshing is basically the same as that for any other processing stage: One must standardize all other processing steps leading to the final product and do the threshing by the normal (local) method and by an optimal method which will give maximum yield of undamaged grain.

As with estimating loss with the straw (threshing loss 1 above), the estimator selects random bundles of cut grain. These are randomly divided into two lots of approximately equal weight. The methodology consists essentially of weighing initially and at the end to compare the traditional (or any other processing procedure) with a processing procedure that gives 100% recovery. Lot 1 is threshed in the manner under evaluation. This may include a final hand-stripping, depending on local custom. The threshed grain, including dry hand-stripped, is bulked.

Lot 2 is hand-stripped carefully and bulked. (Note: Subsamples of each lot may be taken if laboratory equipment is available.) The separate samples are processed carefully to avoid loss or damage through the locally used processing system (cleaning, parboiling, drying, or milling) if this is a batch system in which the samples can retain their identity. The products are then analyzed for broken grains and damaged grains. This is especially important for rice, which is desired as a whole grain, and grains such as red sorghum which undergo a two-stage grinding system wherein bran or husk is first removed from the whole grain before grinding.

If local labor is available, separation of whole from broken grain may be performed by the local method (eg, hand-winnowing): The out-turn of whole grain is calculated and the results for threshing (by one or more local methods) compared with those for hand-stripping.

If the identity of the samples would be lost by processing through the local system (large dryers or large continuous mills), then subsamples should be taken and processed in the laboratory.

MAIZE SHELLING LOSS: Loss on Cob or Core

The method used is basically the same as for threshing: Random samples of cobs are taken and the grain is shelled by the method under test. All the grain is collected and weighed and a sample taken (sample 1). The grains left on the spent cobs are hand-stripped and weighed and a sample taken (sample 2). Moisture content of the two samples of grain is measured with a moisture meter and, if necessary, an adjustment made to the weights. The percentage ratio of the hand-stripped grains to the total is the percent loss. The two portions of grain must be kept separate for the next loss assessment, grain damage.

Losses of insect-damaged, mold-damaged, or stored grain may be different from the losses without such added factors. It is therefore necessary to define the situations being measured and the condition of the grain. For example, losses during the shelling of maize may actually be due to the release of frass (insect chewings, excreta, cast skins, insects and insect fragments) at the time of the shelling process, or the intentional removal of weevils or musty grains (see next section).

MAIZE SHELLING: Grain Damage

Many mechanical and hand shellers cause damage to the maize kernels which can result in a loss of food.

Shelled grain from the previous loss assessment, but *not* the hand-stripped material, is sampled and a representative subsample of at least 200 grains obtained. These grains are examined visually for cracks and scratches, and the number of damaged grains counted and the total expressed as a percentage. It is important *not* to count insect-damaged, mildewed, or shrivelled grains, only damage caused by the sheller. To check this, a parallel sample of cobs should be carefully stripped by hand and at least 200 grain samples also examined. An example of the use of these methods is given in Fig. 7.

DRYING LOSS ASSESSMENT: Loss by Damage

In this section the grain under consideration will be raw paddy rice, though the methodology can be applied in principle if not in detail to other grains and to parboiled paddy. The method is based on that used by a TPI team in

Malaysia and was used to compare three drying methods: 1) yard (sun), 2) batch (Lister), and 3) continuous.

1. Yard (Sun) Drying.

The method for dryer-induced losses based on a laboratory milling operation may be performed in a mill yard, on the highway, or in the farmyard.

(a) Method for Use in a Rice Milling Laboratory on Small Samples

As the bags of one variety of paddy arrive at the yard, they are sampled (see Chapter IV) and blended. The composite or bulked sample (of about 1 to 1.5 kg) is then dried carefully.[4] Meanwhile the paddy will be dried in the usual way

---

[4]"Carefully" means dried in a laboratory dryer with forced air convection at 1.5° to 2°C above ambient air so as to bring the samples to an equilibrium moisture constant (ie, about 14%) in not less than 36 hr.

Maize Shelling--Loss on Cob and Damage Assessment

Assessor's name ___________________________ Date completed ____________

No. of cobs sampled ___________________

Variety ____________________________ Source _____________________

(1) Shelling: Operator's name(s) _________________________________

    Total weight of grain shelled: 5.25 kg

    Moisture content (by meter):   12.5%

    hence weight grain converted to 15% moisture content

$$5.250 \times \frac{100 - 12.5}{100 - 15.0} = 5.40 \text{ kg}$$

(2) Hand-stripping: Operator's name(s) ___________________________

    Total weight of grain stripped: 0.750 kg

    Moisture content (by meter):   12.0%

    hence weight hand-stripped grain, 15% moisture content

$$750 \times \frac{100 - 12.0}{100 - 15.0} = 0.776 \text{ kg}$$

    Hence loss on cob is $\frac{0.776}{5.404} \times 100 = 14.3\%$

(3) Damage: Operator's name(s) __________________________________

    Total grains sampled at (1):   200

    Number of grains with insect, mold damage   25

    Number of grains with sheller damage   2

    Sheller damage: $\frac{2}{200} = 1\%$

Figure 7

and, when dry, rebagged for storage prior to milling; a further sample of about 1 to 1.5 kg is then taken. The two samples (before and after drying) are placed in cloth bags and, as soon as possible after sampling, are dried carefully[4] down to approximately the same level of moisture. A small flatbed dryer with a flow of air only slightly (1.5 °C) above ambient is suitable. Drying to around 14% moisture content should take 6 to 12 hr. After a further three to five days to equilibrate (stabilize), the samples are checked for the exact moisture content and milled.

The best procedure is to use a standard laboratory mill (huller plus cone). Each process should be done in a standard way and in accordance with manufacturer's instructions. The rice will be separated from husk and bran in the laboratory mill. Whole and broken grain proportions are then measured by separating on a hand trier (indented tray) or a small rotary trier (indented cylinder) and weighing.

(b) Method for Use in Mills

If a laboratory mill for small samples is not available or if the data are required for mill use, the following procedures can be used: Large samples (1 to 2 kg) are taken from representative bags being emptied onto the drying yard, so that the total bulked sample weighs at least 25 kg. This sample is then dried carefully[4] in a small batch dryer (as above). A sizable sample of the dried paddy from the yard is also obtained and the two samples dried and equilibrated as for small samples. If parboiling is customary, it should now be performed in a standard manner, suitable to the variety and district. The samples are then milled in a small commercial mill of local type (Engleberg, cone, "modern") and the total product collected. Many small mills that operate on a toll basis are suitable for this purpose. The product is separated into whole and broken grains. If possible this should be done on a separator (some small mills have these and will provide the product fractions already separated). Alternatively, local labor may hand-winnow to separate. The fractions are weighed and the out-turn of whole grain calculated as before (a).

Note: While it may be inconvenient to deal with large samples, use of a commercial rather than laboratory milling system ensures that the results are directly applicable to the local situation.

2. Batch Dryer.

Samples are taken from at least four places near the top and four near the bottom of the drying bin with good distribution across the bin area. Samples must be taken as the paddy is entering the bin (6 to 12 in. from the bottom) and just before the bin is fully charged.

Samples are taken from approximately the same sites as the bin is emptied. Each sample is kept separate in a cloth bag and is *not* blended with the other samples. There will thus be at least eight samples before and eight after drying for each batch. The samples are dried uniformly and carefully[4] on a laboratory dryer as for (a) above, stabilized three to five days, milled on a laboratory mill as in (a) above, and the results tabulated. It is important to compare drying damage on samples from each part of the bin; that at the bottom is frequently overdried and that at the top is frequently re-wetted by transfer of moisture from below, with consequent high breakage during subsequent milling.

The mean figures for brokens for input and for the batch-dried paddy indicate the average damage caused by the drying process. As a guide to maloperation, the differences between brokens obtained from samples of dried paddy from different parts of the bin are important; the mean figures for a whole dryer are not.

3. Continuous Dryer.

With a continuous dryer, sampling of input and output is performed periodically. Samples (1 kg)[5] should be taken every 15 min over a period of at least 1.5 hr. Larger dryer output may require larger samples. If input is varying, sample the same grain in and out of the dryer.

As with batch dryers, it is better if the samples are kept separate. Samples in cloth bags are placed, as soon as possible, in the laboratory dryer (see 1.a). When dried to 14-16% moisture, the samples are kept for three to five days before laboratory milling. The proportion of broken grains should be constant if the wet paddy is of constant quality and the dryer is running consistently; the difference between the mean figures for input and output samples gives a measure of the damage caused during drying.

GRINDING LOSS AS BRAN: Comparative Assessment by Weight

Grains such as wheat, maize, and sorghum may be ground in stone mills, in mortars, or in steel plate or steel roller mills. If the objective is not only to provide a flour or meal but to remove bran, the optimum milling will remove all the bran and leave all the endosperm (inner part) of the grain as flour. The separation of bran from flour is usually done periodically during the grinding; sieves of cloth are frequently used. Winnowing (air classification or purification) may also be used. The bran and other offals will usually be used for animal feed. The problem in assessing the yield of desired product (flour) is that of comparative weighing of various mill fractions over measured time periods. Quality of flour (eg, amount of bran) also may be a factor.

Standard procedures have been evolved for milling wheat on an experimental mill, but this equipment is extremely expensive and of little use for other grains. The methods proposed below may be used to compare the yields of acceptable flour derived from different varieties of the grain or to compare the performance of different operators, and to obtain information on other factors.

1. Comparative Measurement of Milling Yield by Variety

The method selected for milling must be that which is used locally. The ultimate test is milling yield; whatever losses occur must be measured by a standardized procedure.

A number of different operators (eg, women if they are the traditional operators) are required, each with a mill (querns or hand-cranked plate-mills)

---

[5]Appropriately larger samples must be taken if a small commercial-type mill rather than a laboratory unit is to be used, ie, sample size must be matched to test equipment.

of the same type and size.

A portion (about 5 kg) of each variety is given to each operator. Each sample is then milled by sieving or winnowing the product to obtain a flour or meal considered by the operator to be of the usual standard desired in the community. The total weights of grain, flour, and bran are weighed, samples are taken in sealed bottles for laboratory moisture content measurement by oven-drying; and the weights are converted to 15% moisture content basis (or dry weight basis).

$$\frac{\text{Weight flour (15\%)}}{\text{Weight grain (15\%)}} = \text{extraction rate (milling yield).}$$

The average of the milling yield for any given variety obtained from different operators is calculated. Provided that the operator yields for each variety are similar, the method will give an indication of practically attainable milling yield. This same procedure can be run on a commercial mill.

2. Comparison of Operators

With the above procedure (1), a series of milling yields is obtained for a given variety of the grain for a number of operators. If the products obtained were all acceptable to users, the operator attaining the highest yield can be employed to improve the communities' out-turns of edible flour or meal.

3. Comparison of Mills

The procedure of (1) is followed with any one variety to compare the milling yields (extraction rates) for a series of mills.

4. Insect Damage

A constant volume of each grain sample is weighed and milled by a standard milling process and input-to-output of food and nonfood product measured. Insect-damaged grain will give a lower yield of flour than undamaged grain.

RICE MILLING LOSSES

There are many different milling systems in use, but these may be classified as being either one- or two-stage, and either batch or continuous. In the first, the hulling and polishing are carried out in one machine; in the second, separately.

One-Stage Batch Processing (eg, Engleberg Type Huller)

The bag of dried paddy to be processed is sampled and the sample of about 0.5 kg placed in a sealed bottle or plastic bag. The bag of grain is weighed and the moisture content of the grain measured. The paddy is then processed through the huller and the product collected in the customary way. A representative sample of the product is taken. Subsamples (100 g) of the input paddy are then milled on a laboratory mill. The product is separated into husk, bran, and polished rice, and the rice is separated on a hand trier (indented tray) or a small rotary trier (indented cylinder) into wholes, halves, and points. The sample of mill product is separated likewise. The relative proportions of whole grains and total grain are compared; the efficiency of the commercial mill can then be related to that of the optimal laboratory mill and the relative loss calculated.

One-Stage Continuous Processing

As the paddy flows from the hopper or storage bin into the hopper of the huller, a sample (about 100 g) is taken every minute for 10 min. A sample of the product flowing from the output side of the huller is sampled, again a sample of 100 g is taken every minute, beginning about 0.5 min after the first input sample has been taken. The two bulked samples (labeled "in" and "out") are taken to a laboratory and there analyzed by the same procedure as for the batch process.

Two-Stage Continuous Processing

As typical of this system, the "modern" rice mill consists of rubber roll shellers and a series of cone polishers with, perhaps, a finishing brush polisher. Separations are carried out at each stage and after each polishing (usually at least two, frequently four). Skilled operators judge visually the product quality at each stage and also the effectiveness of the separation of product from by-product. Quantitative estimates of machine effectiveness may be measured by sampling on the input and output sides of any machine or battery of machines, processing the input sample by a standard optimized laboratory method, and comparing products for yield (out-turn) and quality (percent of whole grain).

Hulling (suggested basis for a method)

Many mills have two hullers in parallel and some will have a "return huller" for the 10% or so of paddy unhulled in the first pass. It will not be possible to sample the whole product of the huller system, as the material passing back to the return huller has already been through the first huller unit and has been separated from brown rice and husk. Samples must therefore be taken at the entry and exit to each individual machine; if the mill possesses three hullers, each must be sampled separately.

Representative samples (250 g) are taken from the flow of paddy to the huller on a regular basis (eg, every minute), and from the product as it flows to the first separator (likewise every minute) for about 10 min. It is important to obtain a truly representative sample of product; once it has reached the chute leading down to the separator some separation can occur. If possible, the sample should be taken immediately below the rolls.

The well-mixed samples are subsampled for triplicate laboratory testing; the paddy is milled in a laboratory sheller.

The products from the plant and laboratory mills are then examined quantitatively. The ratio of weights of total brown rice gives a measure of the effectiveness of the hulling attained in the plant compared to that in the laboratory. More important is the comparison of the ratio of weights of broken to total grains of brown rice. If the plant huller is giving a higher proportion of broken grains, then wear or a wrong setting on the rollers should be suspected.

Polishing (whitening)

Whether one is endeavoring to measure losses over the whole polishing system or for each machine, the method to be used will be the same: As with other unit operations, samples are taken of the feed to a machine or series of machines and of the product therefrom. The sample of brown rice should be milled carefully in the laboratory to the same degree of milling as that of the machine(s) in the mill. The out-turns of whole grain are measured and compared and the loss in the mill assessed.

Note: Whether it is, in fact, possible to set up such a loss evaluation system remains to be seen. The principal difficulty lies in using a laboratory polisher in one pass to give the same degree of milling as the battery of polishers in the plant and yet also give minimum breakage.

VI. STANDARD MEASUREMENT TECHNIQUES

A. Preamble to the Methodology

K. L. Harris and C. J. Lindblad

Definitions[6]

There is a need to define certain terms and concepts before proceeding to the working methodology.

Losses

This effort deals with removal of food grains from the direct human food chain which, especially in developing countries, is the fundamental energy (calorie) basis of the human diet. The rice weevil consumes rice when living in the kernel. If the kernel is weighed before and after it is bored, it will have lost weight. If the larva or adult is still present when the kernel is eaten, less weight is lost. No consideration is given to a proportional change, if any, in protein accompanying the feeding. The relevance of the insect presence depends on its fate. If it is cleaned out it is loss; if it remains as food, it is weighed as food. Whether insects are eaten or whether the frass is sifted out or falls from bagged grain is sometimes fortuitous, sometimes purposeful. It varies with the season, with the culture, with hunger, or plenty. While the decision to eat may be more socioeconomic than scientific, use or nonuse as food in the specific situation is the controlling factor in these procedures.

Pilferage

In this manual pilferage is not considered to be a loss. It is a transfer of ownership as is spillage when it is used as sweepings in lieu of, or in addition to, wages.

Fungal Damage

It is anticipated that the quantification of weight loss when the loss is due to fungal damage will depend on local practices in the use of the damaged material. People accept or reject damaged kernels as local custom and hunger dictate. One purpose of this manual is to set forth standardized procedures so

---

[6]See also Chapter II, Section A.

that measurements in one country can be compared with measurements made elsewhere. Therefore, in each situation acceptance-rejection limits should be defined in terms of a widely used language. Despite such difficulty, judgment limits based on information obtained from interviews must be quantified.

Processing Losses

Grain removed from the direct human food chain is a loss. Thus milling losses that become animal feed would appear as a loss although reentering down the pipeline with a reduced calorie and, perhaps, improved nutrition input. This "feed" as against "food" use needs to be recognized and described in any situation where it is a factor.

Postharvest

This manual generally accepts Bourne's (1) definition of postharvest as the point at which grain, separated from the plant stalks or root, is bundled for field drying or placed in a container in which it is moved or held, or both. It can extend earlier, however, to include the time during which the mature crop is held in the field for storage or drying.

Household

This manual does not cover losses in food after it reaches the point where it is being prepared for cooking or for direct consumption, even though there can be serious losses in the hands of the ultimate user. In the United States, for example, this may be the most important site. However, estimates and prevention of these losses are so dominated by cultural habits and preferences that in-depth anthropological inputs are required which are not usually within grain loss reduction biology-technology.

Separation From Other Factors

This report anticipates that grain losses will be considered in isolation from other food-availability factors in the areas studied. It is proposed that there is no present need for guidelines that include such sophisticated concepts as how the availability of fish and meat influences the losses, and need to control losses, in staple grains.

Rapid Laboratory-Type Procedures

None of the shortcut tests such as presence of numbers of adult insects, amount of frass, or insect emergence holes are sufficiently accurate when used alone for anything more than loose approximations. "Loss" should be a measurement of actual grain substance removed from the food chain. Techniques for basic statistical concepts are covered in a separate section.

How to quantify losses has been the subject of detailed investigations by the Tropical Stored Products Centre, England, and has been assessed by the Group for Assistance on Systems Relating to Grain After Harvest.[7] Papers

---

[7]The acronym GASGA now stands for Group for Assistance on Storage of Grain After Harvest.

listed in the Bibliography at the end of this section give a definitive appraisal of these losses. From these review papers, from the original published material, from discussions with acknowledged experts, and from first-hand field and laboratory experience come the following conclusions on techniques for measuring losses:

All of the U.S. Food and Drug Administration-generated procedures are too time-consuming, require a laboratory setting, require difficult-to-standardize judgments, are on too small a sample size, or have too variable a relation to grain weight loss to warrant use in determining grain losses. These are the exit hole test, acid fuchsin egg-plug test, berberine sulfate fluorescent stain egg plug test, gelatinzation with sodium hydroxide, and examination for internal insects. Radiographic (X-ray) examinations require expensive laboratory-based apparatus, and are time-consuming and difficult to standardize. The Ashman-Simon Infestation Detector has similar liabilities.

Examinations for insects on the surface of the grain, weighing insect frass (dust from insect chewings and excrement), and various procedures to visually detect damaged grains and count and/or weigh them have been given field trials in developing countries. There is a positive correlation between damage, insects, and frass with some loss quantifications possible and the 1970 IBRD report suggests their use in making *rapid* assessments.

Some confusion exists concerning the application of these procedures in quantifying actual losses. Their use in actual test situations and positive correlations to weight losses have been taken by some to indicate a practical degree of precision to routinely determine weight losses. Such is not the case. They cannot be so used unless the biological and physical characteristics of each assessment situation are completely understood. If lots of grain have the same histories, then their frass-to-loss relations will be similar and may be used to survey them all on a comparative basis. However, if some have been moved (and frass is lost), or some have lesser grain borers (produce much frass), or some have weevils that make exit holes and some have moths that hold their frass in webbing, or the surface insects have been removed from some lots and not others, then any standardization between lots, regions, grains, and countries becomes a new scientific investigation, not subject to rapid comparisons.

However, all of these procedures are of value in a rapid visual and discussion appraisal of a situation to come to a personal judgment. Their precision as indicators of actual losses depends on the expertise of the user. This is discussed in Chapter II, Section D.

Rapid Judgment-Based Procedures

"Guesstimates"

As these estimates with some facts by knowledgeable persons have discovered immediate and urgent needs that could not be met in any other way, they have served many purposes. However, as they have been simple guesses or preconceived opinions for special purposes, they have no validity as determiners of losses. True guesstimates have a valid role in reaching rapid judgments that may suffice for some purposes or precede more accurate evaluations.

Biased Estimates

Although not germane to the present effort, the practical effect of many of the biased figures should not be underestimated. Many have been used to draw forth budgetary support for grain storage and marketing research, build storage structures of sometimes useful value, draw international attention to sometimes real and sometimes imaginary needs, and build local and national stockpiles that have both fed people and wasted grain to the ravages of biological and physical factors.

Traditional Local Estimates

These are especially useful in getting one's bearings on local situations. Interviews should not be passed over lightly. They need to be done with care, as discussed elsewhere in this manual, assessing the point of view and biases of the giver of information, what the figures are based on, and local meanings of such basic terms as "loss" and "percent."

When reinforced by on-site observations or measurements, such estimates may be especially useful in obtaining a picture of local conditions, extrapolating to larger areas, and seeking out specific examples and situations. There are times when local people can make quite accurate comparisons between conditions found in grain as it goes into and is taken out of storage and on actual wastage to insects, birds, and rodents.

On-Site Expert Judgments

While this type of rapid appraisal can be used only by experts to assess percentage or weight losses, its use should not be underestimated.

In making such judgments, one needs to consider how local conditions affect the physical and biological potential for losses. For example, transport in damaged bags or makeshift wagons with visible spillage indicates an obvious loss situation.

Dry conditions spell trouble for insects. At 12% moisture or less, grain insects have a more difficult time feeding and reproducing. By 10% there are serious living problems, and if there is evidence of an arid 6 or 8%, then grain losses to insects are minimal.

Absence of visible insects or damage after six or eight weeks of storage is a good indication that there will be few insects for the next few months also.

The habits of many rodents are well known. Whether stores are open or closed to them, and whether harborages or needed water are available can be readily ascertained.

Losses to rats can be predicted from the nature of the local ecological system. The problem may be more difficult with mice and other small rodents.

Short-term storage, good sound bagging, well-constructed transport vehicles, strict weigh-in/weigh-out control with accompanying records, the use of insect, rodent, bird, and fungal control procedures, and low temperatures all point to minimal losses. Low or high temperature can be of overriding importance. Rice harvested in September in temperate climates may go into natural cold storage before insects make even a minimal start. Grain held under metal roofs or in bags in the sun at over 55°C will have no active insect losses.

On the other hand, while high moisture, active insect, rodent, and bird depredations, and visible mold or heating from microorganisms clearly indicates trouble and potentially heavy losses, the extent of the losses is determined with difficulty by even an expert.

Production and Consumption Figures

Production and consumption figures have often been suggested as a means of assessing losses, the difference between what is produced and what is consumed being loss. Unfortunately, accurate figures at either end of the system are available only in the most sophisticated and developed situations, and the approach is of small practical value in many developing nations and local developing-country locations.

Standardization

Moisture

Changes in volume and weight due to moisture need to be explained. Grain harvested at 21% moisture dried to 15% by mechanical means or aeration has lost weight but not food value.

Measurement of moisture changes requires the use of meters or drying ovens. Weight changes need to be determined by sensitive devices. Use of moisture meters and scales or balances requires such devices and a degree of expertise in their use that may necessitate some basic training. Moisture meters are discussed in Appendix C.

Accuracy

Overall statistical concepts are presented in Chapter IV. It seems reasonably safe to anticipate that 75% confidence limits of ± 5% would, for the present, be as much, or perhaps more, than can be generally expected. However, as yet, there is no fixed gauge as to what constitutes reasonable accuracy. The amount of method variation that may be expected to occur in different commodities, ecological zones, parts of the harvest-to-consumer pipeline, and types of damage by different individual or mixed types of losses are subjects that require clarification in and before any survey appraisal. The first field appraisal should bear these and other factors in mind, particularly as the desired confidence limits influence the duration and expense of the assessment.

Literature Cited

1. BOURNE, M. C. Post harvest food losses — the neglected dimension in increasing the world food supply. Cornell International Agriculture Mimeograph 53 (1977).

Bibliography

ADAMS, J. M. Storage loss assessment techniques, a biologist's view. Trop. Stored Prod. Centre (1972).

ADAMS, J. M. Report on post harvest loss assessment in durable produce, with particular reference to methology. Trop. Stored Prod. Centre (1976).

ADAMS, J. M. A guide to the objective and reliable estimation of food losses in small scale farmer storage. Trop. Stored Prod. Inf. 32 (1976).

ANONYMOUS. GASGA Seminar on Methodology of Evaluating Grain Storage Losses. Trop. Stored. Prod. Centre (1976).
HARRIS, K. L. Evaluation of grain storage losses. Report of the International Bank for Reconstruction and Development (1970).

CHAPTER VI

B. Losses Caused by Insects, Mites, and Microorganisms

J. M. Adams and G. G. M. Schulten

Insects are a major cause of postharvest grain losses. By boring within the kernels and feeding on the surfaces, they remove food, selectively consume nutritive components, encourage higher moisture in the grain, and promote the development of microorganisms.

Methods for detection of internal insects have been summarized earlier in this chapter. Methods given in this section are for determination of losses to the grain itself and are of three types:

1. Determination of the weight of a measured volume of grain (see Methods A and B1). In this case the loss in weight in samples taken over a known time period is a reflection of losses caused by insects or microorganisms, or other factors. Judgment as to cause of the loss is a second and necessary step in the process.

2. Separation of damaged and sound kernels and determination of their comparative weights calculated in terms of the whole sample (see Method B2).

(In both 1 and 2 above, it is usually necessary to obtain a baseline sample of the condition of the grain at the beginning of the test period or to conduct tests to estimate the baseline condition in order to determine the real losses at that point in the pipeline.)

3. Determination of the percentage insect-damaged grain and its conversion into a weight loss using a multiplication factor (see Method B3). (This method also gives an approximate figure for use in preliminary surveys.)

Methodology

Sieving

In all of the methods, prior to analysis, the grain sample should be sieved or winnowed, or both, to remove dust and insects. Use the sieve and sieving/cleaning technique commonly used by the local farm/merchant/consumer for removal of such fractions that would be normally discarded as inedible prior to further processing.

Determination of the Original Condition of the Grain

Since the weight-to-volume method is based on differing weights for different levels of loss, it is necessary to obtain a baseline point, by sample or calculation, from which it is possible to compare all future measurements. This baseline needs to be in the form of a curve covering all of the grain/moisture conditions to be found in the particular grain situation because some grain volumes change significantly, and most often regularly, at varying moisture contents.

The curve is obtained from analysis and calculation of a baseline sample. Determination of the baseline condition is essential so as to have a fixed reference point with which to compare losses incurred during storage. If it is not possible to obtain this sample until after storage or the process under study has already begun, a visibly undamaged sample should be taken and analyzed as early as possible. This should be split into three replicate subsamples and the measurement required by the appropriate methods 1, 2, or 3 applied to each subsample. Each subsample should then be placed in a jar covered with muslin, to prevent insects entering or leaving, and kept for four weeks. At the end of this period, the jars should be examined for insects and damage. If there is no damage in any jar, then all three replicates can be used to calculate a value. If there is damage in one, this must be discarded; if two have damage, both are discarded; and if there is damage in all three, then take the sample(s) with 5% or less damaged kernels. If the damage is above 5%, assistance will be needed from an expert in determining the appropriate correction factor.

Method for Baseline Determination

A sample of approximately 5 kg is either taken from every farmer's store if they are being treated as individual case studies or, if there are distinct grain varieties under study, a representative sample of at least 5 kg is taken for each variety, assuming that they are fairly homogeneous. If any of the varieties is not uniform (does not have a standard weight-to-volume variation with changes in moisture due to intravarietal variations of the local grain(s)), then either each lot of stored grain must be treated individually or expert advice must be sought.

This large sample is sieved in the laboratory. The bulk sample is subdivided into five replicate subsamples. The moisture content of a representative subsample is measured. The range of moisture content which might be expected in the field over the storage season is determined either from locally available data or by approximation (a normal range that fulfills most purposes is 8-18%, depending on climatic conditions). The weight/volume relationship is taken over the range as follows: the range is broken down into five equal steps, eg, if it is 10-18%, this will be 10, 12, 14, 16, 18. If small, perhaps 1%, steps such as from 8-12%, this will be 8, 9, 10, 11, 12%. One subsample will have a moisture content near to one of these figures and the moisture contents of the other subsamples will have to be changed either by drying or wetting, as follows, to cover the range.

Drying down to a moisture content. This should be done with the grain in a shallow layer either in a warm, dry place with a current of air passing over it but protected from insect attack or, preferably, in a ventilated oven in shallow trays at a temperature not exceeding 35°C. Its moisture content should be checked at regular intervals by allowing a sample to cool and measuring its approximate water content. When it has reached the required moisture content, it should be placed in a sealed container to cool and the moisture content should be measured accurately. As a rough guide, a small sample of known weight can be placed on a dish in the oven and its loss in weight checked.

Wetting up to a moisture content. This requires addition of a calculated weight of water to the grain to bring it up to a required moisture content. The weight of water required is given by the formula:

Weight of water to be added (g) = weight of grain

$$\times \frac{\text{Required \% moisture content} - \text{initial \% moisture content}}{100 - \text{required \% moisture content}}$$

For example, if we have a subsample of 1,000 g of grain at 12% moisture content and require it to be at 16% moisture content, the calculation is:

$$\text{Weight of water} = 1000 \frac{16-12}{100-16} = 1000 \frac{4}{84} = 47.6 \text{ g.}$$

This can be weighed out or, since 1 g of water occupies 1 ml, it can be measured out as a volume. Water is added to the grain in a sealed container with sufficient headspace for mixing, and mixed well. It is left for two weeks to condition, but vigorously shaken daily. For moisture contents over 16%, the container should be kept at 5°-10°C in a refrigerator to discourage mold growth. At the end of the conditioning period, an accurate moisture content is determined for each subsample.

There are now five subsamples of grain at different moisture content for each variety. For each subsample the weight that occupies the volume measure (test weight container) should be determined by filling the container (see Fig. 8) according to the instructions provided with the apparatus and then pouring out the contents and weighing it to the nearest 0.1 g. This should be done three times for each subsample and a mean result obtained.

There will now be five mean weights for each variety at five accurately measured moisture contents. Each of these weights should then be converted to dry weight as follows:

$$\text{Dry weight} = \text{weight of grain} \times \frac{100 - \text{\% moisture content}}{100}$$

For example, if the volume of grain in the test weight container weighed 800 g and had a moisture content of 15%, then its dry weight is:

$$\text{Dry weight} = 800 \times \frac{100 - 15}{100} = 800 \times \frac{85}{100} = 680 \text{ g.}$$

This is done for all subsamples so as to obtain a set of dry weights for each moisture content. A graph is now drawn of the dry weight against the moisture content, for example:

% m.c.	10.2	12	13.9	16	17.8
Dry wt.	700	680	650	620	600

From this a reference line can be plotted of dry weights as determined by measuring the actual moisture content and test weight at the time a test is made. This graph is then used throughout the rest of the sampling period to represent the dry weight of sample at any moisture content as if it had not been damaged in store.

A curve must be made for each variety or area-cultural situation (see Fig. 9).

Loss Measurement Procedures

METHOD A — Standard Volume/Weight Method for Damage by Insects and Microorganisms

After preliminary laboratory work for the baseline figure, the measurements can be made in the field or laboratory.

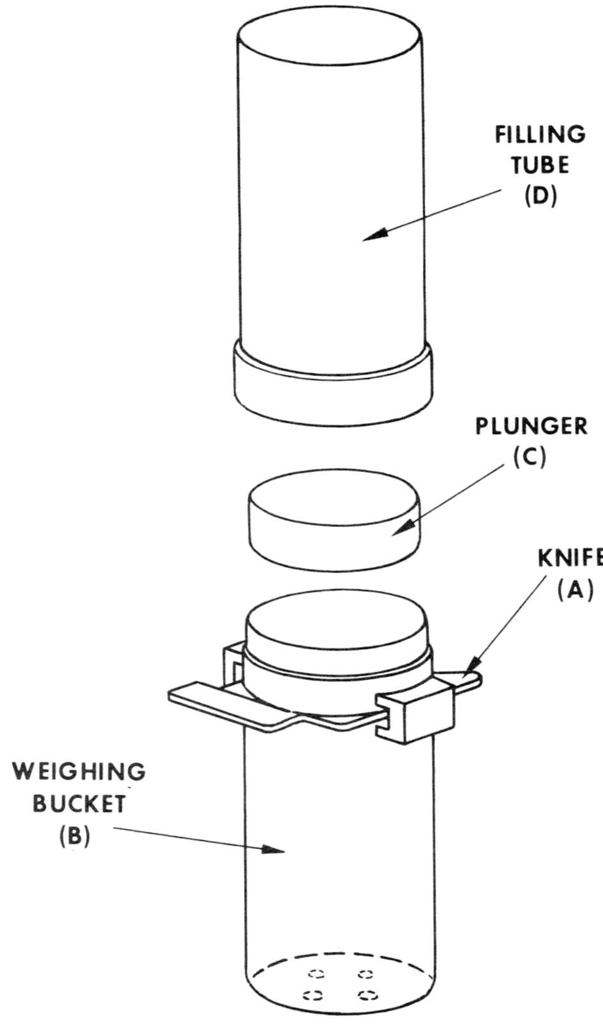

Fig. 8. Test weight apparatus. (A complete chondrometer using this device is available from Simplex Seed Sorters Ltd., Rainham Rd. S., Dagenham, England.)

Equipment
1. Test weight apparatus for obtaining a standardized volume of grain.
2. Balance, such as a triple beam balance, capable of measuring 1.0-1.5 kg accurate to 0.1 g.
3. A moisture meter capable of measuring to 0.1% and calibrated for the type of grain being measured.
4. A suitable size of grain sieve for the removal of insects, dust, and any other material that would normally be removed prior to further processing.
5. Plastic sample bags and a liquid fumigant such as CCl_4 to retain samples for examination at a later date.

Procedure

A well-mixed sample, taken from the store, is first sieved by a locally appropriate method and the weight of sievings are counted as a loss if they are not

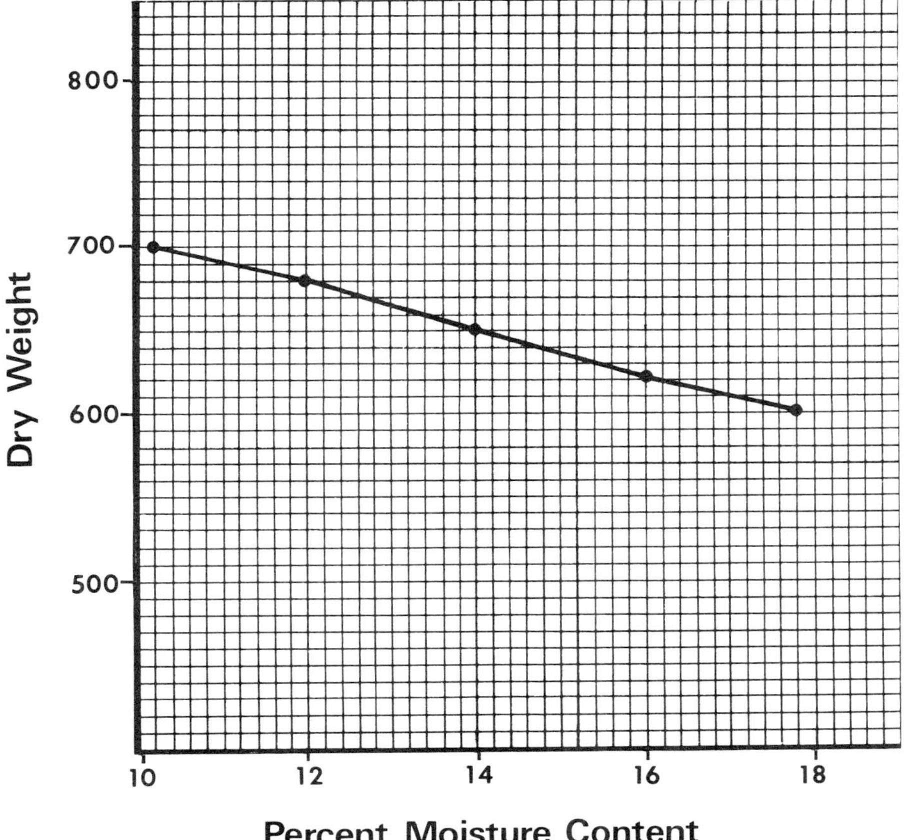

Fig. 9. Example of a standard baseline curve for dry weight of a fixed volume of grain as moisture content changes.

used locally or calculated back to the weight/volume if they are used.

The moisture content is measured.

The weight occupying the volume container is measured. This is repeated three times and a mean taken. This weight is converted to dry weight using the moisture content and formula for dry weight (see derivation of Fig. 9).

The graph is used to find the dry weight of a sample at the same moisture content taken at the time of storage. For example, if the moisture content of the farmer's sample was 12%, then referring to the example, Fig. 9, the dry weight would be 680.

The weight loss in the farmer's sample is then calculated as follows:

$$\% \text{ of weight loss } - \frac{\text{dry wt. from graph } - \text{ dry wt. in sample}}{\text{dry wt. from graph}} \times 100$$

For example, if our farmer's sample at a moisture content of 12% had a dry weight of 600 g, then as the dry weight on the graph for 12% moisture is 680 g, the loss would be:

$$\% \text{ dry weight loss } = \frac{680 - 600 \times 100}{680} = \frac{80 \times 100}{680} = 11.8\%.$$

This is the dry weight loss, which by definition excludes moisture content changes.

Sources of Error

The standardized method of obtaining the volume attempts to eliminate variations in packing, but with grain samples containing very high levels of damage, some of the grains may become crushed and lead to inaccuracies especially with small grains that may be sieved or winnowed out or crushed so that their insect- or microorganism-caused emptiness is not detected. In this case they may have to be picked out and losses otherwise estimated. Conversion factors change in the course of the storage period from high to low, due to increased severity of damage to the already-damaged grains.

The admixture of an insecticidal dust to shelled grain increases friction between grains and will reduce packing and hence the weight per unit volume will be less. Therefore, the weights for treated grain must not be compared with weights obtained for untreated grain.

For paddy, the effect of moisture content on the dry weight occupying a given volume is negligible, so within a range of 5% moisture there is no requirement for a predictive graph.

Rice (as distinct from paddy) would best be measured by out-turn of the mill.

Lumps of, or otherwise webbed-together, grain can add weight. However, if the lumps are picked or sieved out by local custom, they should also be picked out and the kernel loss estimated.

Since little is known about methods for determining losses in insect-damaged millet which, in effect, are hollowed shells, and since no procedures have been satisfactorily described for picking out and weighing of insect-infested millet, this grain presents a real problem not yet resolved by this current methodology.

METHOD B1 — Modified Standard Volume/Weight Method When a Baseline Cannot be Determined

The standard volume/weight method as described under METHOD A is presently the most reliable method of loss determination. There are, however, situations where this method cannot be used without modification. It may also be difficult to obtain reliable moisture content determinations in some cases.

It is often necessary to make loss estimates in the middle of the storage period when no baseline has been previously determined. It also frequently occurs that in a rural area different varieties of grain are grown under different conditions, such as with or without fertilizer, or on poor or good soils. This may affect the size of grains and, consequently, the volume/weight ratio.

Application of insecticide dusts may also affect the settling of the grains in the standard volume and increase the volume occupied by the grain.

Because of these various conditions, a separate baseline may have to be determined for each individual farm or storage situation. This is often impossible to achieve between harvest and storage.

Procedure

The standard volume/weight method should be used but an artificial baseline should be prepared by selecting undamaged samples from the grain present in the store at the time of loss determination. The loss is the difference in weight (expressed as a percentage) between the undamaged and the damaged sample. Conversion for moisture need not be used in this case since the moisture content will be approximately the same.

Experience with this modification of the method is still limited. For maize ears stored with husks, it is possible to select a number of undamaged ears, to shell these, and to use the grains to determine the baseline. With other grains, it may be more difficult to obtain an undamaged sample, especially in cases of heavy insect infestation.

Sources of Error

Unreliable results may be produced if, during selection of undamaged grains from the stored grains, there are hidden internal infestation, preferential feeding and egg deposition by insects in grains of different sizes, and a difference in moisture content.

To overcome the problem caused by hidden infestation, the same procedure for obtaining an undamaged sample as indicated for the normal standard volume/weight method can be followed.

Insects do not often feed or oviposit on grains at random but, depending on species, they may show a preference for smaller or larger grains. There is then the risk that in selecting undamaged grains, a particular grain size may be selected which is less liable to infestation than grains of another size. Grain size obviously affects the volume/weight ratio. When undamaged ears are selected, there is the possibility that smaller ears (with smaller grains) may be unintentionally selected, since smaller ears are less often infested than larger ears due to a better husk protection. The only way to reduce this error is to take the undamaged sample as much at random as possible. In addition, a sample must be taken which is larger than necessary and, after good mixing, only a part of the sample should be used for baseline determination.

When the baseline and field sample are taken from the same part of the storage structure, it is not usually necessary to determine moisture content since differences between the two samples are likely to be small. The weight difference between the two samples represents the actual loss. If there is doubt about the homogeneity of the moisture content of the grains in store, the method for baseline determination should be used.

Insects prefer moist grains rather than dry ones. This behavior may cause the baseline sample to be drier than the field sample. When this is suspected to be the case, the method for baseline determination should be followed. When this is impossible, the error can be reduced as much as possible by taking large random samples.

METHOD B2 — Count and Weigh Method

There are many situations in which a loss estimate is required but where there is only minimal equipment available and the baseline could not be determined before the storage period. In addition, it is sometimes impossible to determine a baseline for the standard volume/weight method because too many grains have been damaged.

This is essentially a method that takes a sample, separates it into undamaged and damaged portions, counts and weighs each, and calculates the percentage weight loss. It assumes that the undamaged portion is totally undamaged.

Used for unshelled and mold-damaged grains, it provides a useful means of estimating loss at moderate infestation levels with a minimum of apparatus.

Equipment

1. Balance with a range of 0.5 g to 1.5 kg accurate to 0.1 g.
2. Tally counter.
3. Plastic bags and a liquid fumigant such as CCl_4 to enable retention of samples.

Procedure

The grains are separated into undamaged and damaged categories, the latter being separated according to cause. Grains in each category are counted and weighed. The resultant data may be substituted in the formula below:

$$\% \text{ weight loss} = \frac{(UNd) - (DNu)}{U(Nd + Nu)} \times 100$$

where U = weight of undamaged grains,
   Nu = number of undamaged grains,
    D = weight of damaged grains,
   Nd = number of damaged grains.

Sample Size

Experience with this method is still limited. A sample size is recommended of 100-1,000 grains. Besides its simplicity, the method has the advantage that damage by different species of insects, such as *Sitophilus, Sitotroga, Ephestia* spp., and *Rhizopertha,* can be measured. The method may also be used to determine damage caused by termites, rodents, and birds.

Sources of Error

Hidden infestation results in an underestimation of loss because grains that have lost weight are included in the undamaged portion. When the grain is heavily damaged, it may become so broken as to lead to counting errors.

At low levels of infestation with the insects selecting larger or otherwise nonrandom grains, the method is not dependable. At very high levels of infestation, kernels may be so destroyed as to be not measurable. For example, in maize ears at low infestation, often only the grains at the top of the ear are damaged because they are incompletely protected by the husks. These grains are often the smallest of the ear. The only recommendation to reduce this error is to take large samples.

Since insects will sometimes select and infest larger kernels, any procedure that compares the individual weights of kernels may result in a negative weight loss finding. The selection of internally infested kernels and their inclusion and weighing as undamaged can also result in negative loss findings unless care is taken to recognize and account for these samples.

A preference of insects for moist grains may confuse the relation between weight loss and damaged grains as well. To reduce a possible error arising from this behavior, the grains could be dried to the same moisture content.

METHOD B3 — Converted Percentage Damage Method (For Use in Field or Laboratory)

This method is suitable for insect damage only and provides a useful estimate for quick appraisal of losses without needing equipment. It can be easily used by unskilled but trained personnel.

When grains are heavily infested, feeding by secondary pests and multiple infestation may disturb the relation exit/weight loss and so lead to an underestimate. Therefore, when possible, it is preferable to determine the conversion factor instead of using those factors indicated below. It will be obvious that the conversion factor can be ascertained in a sample at any time after the sample has been taken as long as the sample is properly stored.

When losses have to be measured in a large number of samples, originating from cereals which were stored for about the same period of time and under similar conditions (eg, regional surveys), at least some samples should be kept for determination of the conversion factor.

Although the converted percentage damage method is liable to the same sources of error as the modified standard volume/weight method and the count and weigh method, it has given very good results in practice.

When earlier-mentioned methods cannot be used, it is recommended to use the converted percentage damage method rather than guessing. With this method, weight losses in cereal grains and pulses are determined in a slightly different way.

Equipment

1. Tally counter.
2. Plastic bags and a liquid fumigant such as CCl₄ to fumigate samples when determinations are done at a later date.

Procedure

The number of damaged grains is counted in the sample and expressed as a percentage. This percentage damage is converted into weight loss by means of approximate conversion factors as indicated below. This factor can be determined separately for each individual situation or established factors can be used. This loss determination is only applicable when the damage has been largely done by insects which leave a clear exit hole in the grain *(Sitophilus, Sitotroga,* and Bruchidae).

CEREAL GRAINS

A random sample of 100-1,000 grains is taken and the number of bored grains is counted. This can be done immediately or within a few days after sampling. When there are too many samples to be counted, it is recommended to store each sample in a sealed plastic bag to which some liquid fumigant has been added.

The percentage of damaged grains is calculated with the following formula:

$$\frac{\text{Number of bored grains}}{\text{Total number of grains counted}} \times 100 = \text{\% bored grains in sample.}$$

This percentage is converted into a percent weight loss by dividing it by the conversion factor (C) or multiplying it by 1/C.

To determine the conversion factor, a random sample of 100-1,000 damaged grains is taken which contains 10% or more damaged grains. The percentage weight loss is determined with the count and weigh method, and the conversion factor is calculated as follows:

$$\frac{\text{Number of bored grains}}{\text{Weight difference in \%}} = \text{conversion factor.}$$

The following conversion factors have been established in practice where the larval stages develop within the grain, eg, *Sitophilus* species, *Sitotroga cerealella:*

Maize (stored as shelled maize
     or as ears without husks)   % bored grains   × 1/8
Maize (stored as ears with husks) % bored grains   × 2/9
Wheat   % bored grains   × 1/2
Sorghum   % bored grains   × 1/4
Paddy   % bored grains   × 1/2
Rice   % bored grains   × 1/2

PULSES

In pulses several well-defined exit holes may be found in one bean or pea. When infestation is not too heavy, it can be assumed that each weevil consumes about the same amount of food for its development. Therefore, in the case of pulses the number of exit holes is counted and not the number of bored beans (peas). For determination of the conversion factor in pulses, the same procedure is followed as for cereal grains but the damaged sample must consist of beans (peas) with one exit hole only. The conversion factor indicates the number of exit holes which equals a weight loss of 1%.

In the field sample, the number of exit holes has to be counted in 100-1,000 beans. This number is divided by the conversion factor and the percentage weight loss obtained.

A known conversion factor for cowpeas when bruchids are the cause of damage is number of exit holes in 1,000 grains divided by 200.

CHAPTER VI

C. Losses in Grain due to Respiration of Grain and Molds and Other Microorganisms

R. A. Saul, with K. L. Harris

A mass of grain can be considered as a living organism that feeds on itself. It is made up of the individual seeds which are hosts to the many microorganisms of fungus, yeasts, and bacteria. It loses or gains moisture depending on its moisture content and the ability of the surrounding air to absorb or release moisture (relative humidity). For example, maize at 12% moisture in air of 75% relative humidity will gain moisture until it reaches 15%. If the grain moisture gets high enough the grain will sprout. At lower moisture levels the seed is essentially dormant and has a very low and rather constant rate of respiration.

Microorganisms can grow under lower moisture levels than grain. They take moisture from the air and use it for their metabolism. Yeasts and bacteria require an atmosphere of 95% relative humidity or higher, while fungus grows under conditions as low as 75% relative humidity.

The rate of growth of the microorganisms is dependent on temperature as well as moisture. Also, the extent of physical damage to the kernel is a factor which influences the rate of growth.

Growth of the microorganisms and the seed is at the expense of the seed dry matter. The rate of growth is reflected in the rate of dry matter loss. When quality is reduced to the degree that the grain is rejected, there is an additional loss of quantity.

Weight losses due to respiration of the grain itself are unimportant until the moisture is so high that serious deterioration by microorganisms occurs. In other words, when there are serious quantitative losses due to respiration, the quality has so deteriorated that total, or kernel by kernel, rejection for food use becomes the dominant factor, not losses in weight due to respiration. At this point, determination of losses involves an appraisal of amounts of grain rejected for food use.

The conclusion must not be reached that if there are no changes in weight that the grain is free of mold toxins. Toxins are a separate matter. When suspected they must be determined by special tests.

Thus, there are two types of losses. One is the loss due to grain being converted by microorganisms to carbon dioxide and water. The other loss occurs when the grain (in its entirety or as individual kernels) is rejected as food. Such rejection can occur because of an obvious discoloration or odor, or because of the more technical knowledge or implication that harmful substances (mycotoxins) are present. In the latter situation, one must determine the amounts of grain rejected for food use.

Any visual survey by locals or outsiders on what an individual rejects or accepts becomes a difficult assessment. It needs an input of all of the principles of measuring subjective values, bearing in mind that bias is eliminated with difficulty and that all elements of bias are probably not completely known.

Loss Measurement by Standard Table Based on Time, Temperature, and Moisture

It is the nature of molds, yeast, and bacteria to reduce organic material to simpler organic compounds or even to its inorganic form. That is, molds decay the grain and, if conditions are favorable for the growth of mold, then they will destroy the grain.

Long before the grain is completely destroyed, it is made useless as food because of the musty odor, discoloration, and possibly formation of toxic substances. In fact, this will occur by the time 1 or 2% of the dry weight has been destroyed.

The rate of loss of dry matter due to mold growth depends on, in order of importance, grain moisture content, temperature, and amount of physical damage to the grain.

Although, as stated earlier, yeasts and bacteria grow at moisture levels lower than those required by grain, a high moisture environment of 95% relative humidity or higher is required for growth. Grain in equilibrium with this relative humidity will be about 22% moisture, depending on the temperature. Rice and maize are often harvested at this moisture content, but most other grains and seed crops are harvested at lower moistures. Mold, however, can grow under these and even somewhat drier conditions. Mold growth stops below conditions of 70% relative humidity. Safe storage moisture content for grain will be below that in equilibrium with 70% relative humidity. Some molds can grow very slowly in grain at temperatures below freezing of water, but at temperatures of 54.5 °C their growth is stopped. Table III shows the rate of dry matter loss in relation to temperature and moisture, and shows how much weight loss may be expected to occur in undamaged grain at given moistures and temperatures. As seen in the Table, grain at 25% moisture and 15.5 °C will lose 0.0312% of dry weight per day. Thus, in 60 days the loss will be: $0.0312 \times 60 = 1.87\%$. By this time the grain will be obviously out of good condition.

TABLE III

**Rate of Dry Matter Loss in Undamaged
Grain as Related to Grain Moisture and Temperature**

Temperature (°C)	% Loss per Day			
	15% m.c.[a]	20% m.c.	25% m.c.	30% m.c.
4.5	0.0003	0.0033	0.0098	0.0173
15.5	0.0010	0.0106	0.0312	0.0553
26.5	0.0034	0.0338	0.0994	0.1766
38.0	0.0101	0.1074	0.3165	0.5622

[a] m.c. = moisture content.

Notes: Oilseeds will not necessarily follow this table. Mechanically field-shelled (combine-shelled) maize will regularly contain approximately 30% damage and Table IV will apply. *Below* 15% moisture-caused losses will be inconsequential.

Damage to the seed coat of a kernel creates a more favorable condition for mold growth. Physical damage is defined as any break or rupture in the seed coat of the grain. Physical damage is associated with shelling or threshing and is also caused by insects and rodents. It can be pronounced in corn mechanically shelled at high moisture levels. Small grains such as wheat and rice would have very low levels of damage due to harvest but insect damage should be considered. Table IV shows the factor by which the rate of loss for undamaged grain in Table III is multiplied to estimate the rate of loss for damaged grain. Thus, if the loss were 1.87% as calculated above and the grain had originally had 10% damaged kernels, then 1.87% must be multiplied by 1.30 and the loss would come to 2.43%.

Tables III and IV apply to the first 1 or 2% loss of dry matter. The rate of loss will increase with time as the molds grow and multiply; however, the grain will generally be rejected as food by the time 2% loss has occurred.

Moldy grain may be unevenly distributed in layers or pockets associated with high moisture from leaks, condensation, and insects. In such cases, it is necessary to measure separately moisture and temperature in these pockets and in nonmoldy portions of the grain.

Loss Measurement by Weighing Damaged and Undamaged Kernels and Calculation of Loss

The sound and moldy kernels should be counted and weighed and the average weight determined.

$$\% \text{ weight loss} = \frac{(\text{UNd}) - (\text{DNu})}{\text{U(Nd + Nu)}} \times 100$$

where U = weight of undamaged grains,
   Nu = number of undamaged grains,
    D = weight of damaged grains,
  Nd = number of damaged grains.

Samples taken from stored grain may contain kernels from portions significantly stratified as to mold and moisture (insects also), and in the calculation of losses it may be necessary to allow the samples to reach a moisture equilibrium before weighing. Internal insect-damaged kernels may be present in both

TABLE IV

**Physical Damage Modifier on Rate
of Dry Matter Loss**

Physical Damage (% by weight)	Modifier
0	1.00
10	1.30
20	1.67
30	2.17

the sound and moldy portions and may need to be considered. Experience has shown that, if as much as 1% is insect infested, infestation will be visible as insect emergence holes when about 500 g of grain is rapidly examined for insects or insect damage. This examination is conducted by passing a small amount of grain at a time across a well-illuminated surface, and rolling or turning the kernels while searching for emergence holes. The 500 g can be examined in about 5 to 10 min. Such an examination should reveal some, but not necessarily all, of the holes.

Loss Measurements by Comparison of Weigh-In and Weigh-Out

Losses will be measured from start of storage until grain is removed from storage. The method to use for measuring loss should be based on changes in unit weight (test weight). As mold destroys dry matter, it will reduce the unit weight of the grain.

To use this method, a baseline for each storage unit needs to be established by sampling the grain when it is put into storage and measuring the unit weight from this sample, which becomes the basis for estimating loss from future samples from that storage.

Respiration-Induced Losses That Result in Grain Being Rejected as Inedible

Any measurement of weight loss due to respiration of microorganisms would be dominated by a quality loss which would make the individual kernels so bad they would be picked out and thrown away (or fed as feed), or the lot would be rejected.

Therefore, the methodology is to determine what is locally not used for food. This needs a survey technique. The survey will measure a level that depends on a subjective measurement that will vary with time, place, and hunger. In surveying, comparative or permanent use of the data requires that a sample or photographic record, or both, be kept of what the rejection levels were during the particular survey.

Experience has shown that grain may be rejected as inedible when there is about a 2% loss in weight due to mold damage. The level at which this occurs is highly variable and subjective. It varies by socioeconomic levels, by local beliefs and customs, by the degree of hunger, the season and what is available, by whether one is a seller or buyer, and by the difference between common practice and a demonstration for the outsider. There is, however, no definitive answer to the problem of how to obtain a realistic appraisal of actual conditions of use.

The appraisal must fit the local use situation as set forth in the following guidelines:

1. Consult with the person regularly making the decision.

2. Take care that the sex of the person interviewed is the same as that of the person regularly making the decision.

3. Take care that age and other social status situation is as practiced.

4. Take care that outside pressures are not applied:
   a. To be more careful.
   b. To be less careful.

    c. To demonstrate special sight or odor skills.

    d. To impress a husband (wife), headman, outsider, etc.

5. Take care that location, light, time of day, and utensils are normal.

6. Consider using internal checks such as replicate samples or repeat samples on other days.

7. If for home use, buying, selling, or market grading, have the total situation appropriate to that decision.

8. Recognize the importance of interviewer-related bias.

    a. Standardize the approach.

    b. Consider the use of one interviewer throughout the survey.

    c. Consider the use of identical sex, age, and size of local individuals throughout the survey.

Additionally, individual and/or local and/or seasonal or yearly standards with national and international criteria suitable for area-to-area, year-to-year, and country-to-country understanding and standardization may be compared but require detailed expert consultation.

As stated earlier, in any level-of-rejection evaluation, it is imperative that the level be in a form that can be preserved in photographic or other standards, so that there will be a record of what the levels were and what was rejected.

This transfer from local decision to technical standard is one for an expert grain grader. Such a person can transfer the subjective criteria to an area-wide survey of the frequency of occurrence.

The standardized grading approach could be undertaken from the beginning with an experienced grain grader who would use the information obtained in the field to establish a standard of grading which could then be used to train the laboratory technicians. If a sound field basis of judgment were established, it could be uniformly and accurately applied by the trained technicians. This would remove that bias which results when relying on a farmer's judgment. It would also reduce interview time and the time to sample the farmer's store. It would be necessary, however, to either have the experienced grain grader present in the field and laboratory during the entire time of the first year's sampling, or to maintain the samples as evaluated by the farmer in a way that the grain grader could use them to establish the rejection standard. The second approach would be preferred since the seasonal effect could be observed and a realistic average would be more easily obtained. From this, then, would develop a grain standard which would allow grading of any sample of grain on the basis of edible or unedible, and therefore an estimate of the loss of grain as food within the area.

CHAPTER VI

D. Rodents

Part 1. General Considerations, Direct Measurement Techniques, and Biological Aspects of Survey Procedures

W. B. Jackson and M. Temme

Food losses to rodents are acknowledged to be great, but quantification of this diversion from human food supplies is less than satisfactory. Literature on rodent depredations to food (both pre- and postharvest) recently has been summarized by Jackson (1). Lack of adequate data and appropriate survey or sampling techniques was recognized as a prime deterrent in obtaining adequate estimates of loss.

Most data of local or national postharvest losses result from bureaucratic guesses. Studies are rarely undertaken, although extrapolations are sometimes attempted. (See Jackson [1] for detailed analysis of this problem.) While many of the figures quoted in government reports may be correct, they usually cannot be documented.

Most of the surveys which have attempted to obtain data must be suspect, such as a "felt loss" survey among Indian stored-grain merchants who reported that monthly losses (from all pests) ranged from 1.7 to 3.75% of their stocks. Another report notes that 1.7% of sacks holding cacao beans in a Nigerian warehouse were opened by rats and "estimated" that 10% of the stored product was damaged. Estimates by different investigators of postharvest losses to rodents in India range from 2.5-5.9% to 25-30%, and even higher, and annual village losses in India were "estimated" from 2.3 to 3.3 metric tons.

A few small-scale studies have provided some statistics. A 1975 study in one godown in India over 11 months showed losses of 1,400 kg of food grains due to 200 rats. Some rodents hoard food, 3 kg having been found in a single burrow; but the time required to amass such a volume is generally not known. Other estimates of burrow hoards have been as high as 15 kg.

Most efforts at rodent-damage assessments have been focused on crops under field conditions; however, even in the most recent summary of methods (2) only sugarcane is cited as having an acceptable survey tool. Suitable techniques for field assessment of damage to rice also have been developed and field tested in the Philippines.

It is evident that one cannot turn to an existing body of knowledge for obtaining an accurate measure of postharvest losses. It is acknowledged that "the usual method of estimation is to blame vertebrate pests for all losses that cannot be accounted for in any other way." FAO, in assessing its role in reducing food losses, indicated that no agreed methodology existed for assessment of losses from pests generally. Present resources available for the necessary engineering, biological, and statistical studies to develop and evaluate procedures in each country were deemed inadequate. However, GASGA and several FAO projects are now devoting effort to this concern. The program at the Vertebrate Pest Center, Karachi, is of particular interest, but no working reports are known to be available.

Field Losses

Postharvest losses often are assumed to start with some manner of storage, though it must be recognized that crops that are shocked or windrowed in the field for drying may well have rodent infestants and that these rodents can cause local damage and then be transported into storage sites. Assessments may be made in the field directly (usually involving a sampling technique) or with indirect procedures.

If comparable fields without rats can be found, weight or volume differences in the ultimate harvest would provide a good estimate of rodent losses — if fungus, insect, bird, or large mammal depredations were not involved or were assessable. Techniques developed for assessing bird damage to maize utilize counts of individual kernels destroyed, the length of kernel rows eaten, or simply the proportion of the ear damaged. Separation of primary from secondary involvement also is necessary. For example, insects or smut may be able to invade the maize ear when the husk has been penetrated by bird or rodent activity.

If the crop is left in sheaves or stacks in the field for a time, serious damage may be caused by rodents. This damage can perhaps be measured by comparing grain losses and contamination in the damaged portions with sheaves and stacks that were protected from rodents.

Threshing yards are known to be sites where considerable rodent damage and loss can occur. Comparing pre-threshing harvest estimates with grain finally used may ascribe losses to the wrong operational sector, however.

Storage Losses

Direct determination of actual losses is one approach, although the total volume of stored products usually cannot be examined due to time, manpower, or financial limitations, so a sampling technique must frequently be used. Obviously, moisture losses and damage from insects, fungi, birds, or other pests must be assessed separately.

Changes in quality of stored food can be important. Loss of germ by selective feeding markedly reduces the value of maize. Urine, fecal, or hair contamination of stores may provide a disease potential (eg, Salmonellosis) and alter the aesthetic valuation, and hence the market price, of the product.

Unlike insects, which often are distributed throughout the grain stores, rodents will be at the periphery of bulk storage and often nonrandomly distributed through bagged or boxed products. This complicates any statistical approach to sampling and assessment. One approach would be to examine all susceptible products by inspecting *each* bag or container incoming and outgoing for rodent damage (and urine by ultraviolet light if this form of contamination is of concern). Contents of each damaged unit would require detailed examination to determine actual loss. The remaining portion may be judged satisfactory for use, convertible to animal food (at lower market prices), or unsuited for any use. Operationally, products stored in certain structures or sections of structures known to be without rodents could be omitted from such routines.

To ascertain rodent damage and contamination in the total contents of bulk storage units, such stores can be sampled around their perimeter to determine incidence of droppings and gnawed kernels, but this is likely to be most difficult because of inaccessibility of this layer.

Sampling schemes extensively used for assessing grain quality, especially in transit, will be satisfactory for determining rodent infestation or contamination only if the period of transit is relatively short, the load is well mixed, and a large active rodent population is not present. Allowing loaded boxcars to stand on a siding for several weeks permits invasion from local populations, but damage is likely to be peripheral and not detectable by probe samplers.

Indirect determinations of losses involve learning the sizes of infesting rodent populations. If the rodent population can be censused or estimated, their daily food consumption (and contamination) could be extrapolated as an estimate of the loss. The techniques used to estimate population size require statistical assumptions that cannot always be met, although some simple techniques that can be utilized to determine population numbers in most storage facilities are described in Chapter VI, Section E.

The now classic techniques used to census rat populations in New York and Baltimore (3, 4) require calibration for each environmental complex of concern. Even so, this may represent the most practical approach. Essentially the rodent activity in evidence (droppings, runways and burrows, gnawed food) is evaluated by one team and the population size estimated on the basis of these signs. After this a second team determines the actual rodent population by intensive trapping. When the population estimates of the first team are in essential agreement with the capture determinations of the second team, the first team continues through the area with sight surveys and consequent population estimates. Unfortunately, this calibration process is lengthy and must be repeated whenever different species or different environments are encountered. Its adaptation to village or godown environments has not been specifically demonstrated, but as long as the areas of rodent activity are discernible, its application should be possible.

Some attempts occur at popularizing estimation of rodent numbers by assuming the rats seen during daylight hours represent a scientific proportion of the total population. Unfortunately, such procedures are without experimental backing. Furthermore, rats with a larger home range and daily need for water may be more rapidly observed than mice that remain hidden within their food supply.

On a limited basis, direct and total counts of a population may be obtained in a circumscribed area and losses estimated by calculating the food eaten by the population. This involves trapping, marking of individual animals, and direct observation. This tends to avoid difficulties with widely varying movement patterns and nonrandom distribution of animals but is very demanding of time. This requires some judgment as to migrations in and out of the area, amount of grain as against refuse eaten, etc.

One traditional estimation technique employs census baiting. By ascribing a given quantity of a placed bait eaten to a rat, the population can be estimated.

However, where high quality food is stored and thus competes with placed baits, the competition and the neophobic responses (of rats) are likely to result in serious underestimates of the actual population. Mice, with very limited home ranges, often cannot be estimated with such a technique when they are infesting food-storage facilities.

If the population has been satisfactorily assessed, an attempt can then be made to estimate the corresponding losses, or at least the losses caused by the predominant species, for it is rare for only one species to be involved.

A minimum estimate can be made by multiplying the daily consumption of an individual by the number of individuals in the population. Consumption is related to the liveweight of the animals. Mean daily consumption varies with the nature of the foodstuff and especially with its nutritive value. For cereals, the following amounts of grain can be used: For *Rattus norvegicus,* 20-25 g, *Mus musculus,* 2.5-3.5, *Mastomys natalensis,* 8-10, and *Bandicota bengalensis,* 9-11.

If no experimental data are available, daily consumption can be estimated at 1/10 of the mean liveweight of the species.

In addition to the grain eaten by rodents, there are partially eaten grains which are unfit for human consumption. Decisions on discarding such grain will vary with the season, with the abundance of any particular harvest, with local and national mores, etc. Thus, losses need to be on the basis of actual discards, not what should be discarded according to aesthetic and health consideration (see Chapter VI, Section C).

One very real concern is for the process of obtaining accurate data. Catching rats and then releasing them (for Lincoln Index estimates) is difficult to explain to a farmer suffering from rodent depredations. Probably such an approach should be reserved for government facilities where research can be conducted without intrusion into personal rights. Yet studies ought to be done in housing units, local godowns, and small shops or markets. Residents and owners must have confidence in the investigator and must be able to see some direct benefit to themselves for their cooperation, such as removal of rats or better storage conditions. Without the full support of local peoples, the data derived from study programs are likely to be another set of "estimates" that are not well grounded.

Pragmatically how *much* damage or loss occurs from rodent infestations is less important than getting to the sanitation, construction, and control techniques that will result in more stored foods being available to people. But to justify and evaluate rodent management programs, cost/benefit ratios have to be determined. Herein lies the reason that such documentation needs to be undertaken.

Summary of the Problems

Each component in handling and transportation of foods following harvest must be evaluated separately.
- In-field losses lend themselves to direct appraisal (weight loss, kernels damaged) and use of sampling techniques.
- Transportation from one site or field to another may enclose rodents within a food supply. Especially if the vehicle is relatively small and the

time great, losses can be of real consequence. Determination of weight loss, especially after damaged or contaminated portions are removed, can be made directly.

- Local storage — either in the home or in local godowns — is the fate of most grain, and these sites are the most vulnerable to substantial losses. Direct measurements (weight/volume) of depredations are most readily done, but interpretations must be integrated with local environmental conditions.
- Bulk storage, because of larger volumes involved, is likely to have less damage proportionately. The ability either to determine numbers of rodents or to assess the damage itself is more limited, however. If the grain is bagged or containerized in some way, damage to specific containers and their contents can be determined. Contamination especially is of concern in bulk storage, since the mixing of a small quantity of contaminated or infested grain with a large quantity of clean product results in a total lot of contaminated product.
- Economic (and aesthetic) thresholds for food damage and contamination need to be established (5). Efforts at sampling become increasingly costly at lower infestation and contamination rates.

Methods-Oriented Summary

The problem of postharvest losses to rodents resolves itself into three aspects: 1) Losses due to the removal of corn, sorghum, and millet in which grain is eaten from cobs, heads, or spears; 2) losses to threshed or shelled grain; and 3) losses caused by contamination in which the contaminated grain is discarded. (Losses due to rejection by the users is discussed in Section C of this Chapter.)

1. Losses to Ears or Heads of Corn, Millet, and Sorghum

Measurements consist of estimating the percentage of grain removed from the heads, shelling, and weighing undamaged heads of the same size, and calculating losses by percent or actual weight loss.

Samples may be taken so as to be representative of the lot as a whole if the damage is distributed throughout the lot. When damage is located in a particular portion of the stack, pile, or windrow, sampling needs to be representative of that situation (see Appendix B) with an estimation of the proportion of the whole that is so affected.

2. Losses to Threshed or Shelled Grain

Problems of sampling bagged or bulk grain are of three types: a) Those in which before and after weights are available or may be obtained; b) those in which bagged grain with and without damage may be weighed and compared; or c) those in which no actual comparative weights may be made of the grain itself. These procedures are amplified below:

a. In many market, transport, and warehousing situations, the grain has been previously weighed. Reweighings will give the amount lost to rodents, if this is the only source of change. This can be a laborious and costly task, however, and usually an estimate must be made using one of the procedures in the following two paragraphs.

b. Comparison of weights of undamaged and damaged bagged grain: Rodents often concentrate their feeding and nesting in fairly well-delineated areas of bagged grain storage. When this is the case, damaged bags may be weighed and compared with the weight of undamaged bags taking appropriate care to obtain representative samples of the bags if weights before loss are not available. When the individual bags have already been weighed, direct and actual losses may be readily obtained.

c. Overall losses to grain in storage: Most often serious rodent losses occur in relatively long-term storage or in a long-established marketing or warehousing situation where grain is present under a stabilized pattern. With long-term storage, local rodents may be found out of the store, moving in for feeding and subsequent habitation. They will live in the stored grain if undisturbed and if water is nearby. Rodents in markets where there is a permanent supply of grain moving in and out of the storage will usually be living nearby, in holes in or under the floor, between walls, or in burrows, moving into the grain for food, and to nearby sewers, drains, or sinks for water. In these cases, losses involved are estimations of the rodent population, and the food loss is calculated on the basis of the number of rodents $\times$ time $\times$ food consumption.

Some simple methods suitable for general use of rodent population estimation are given in Chapter VI, Section E. Rodents, however, are known for their diverse feeding habits and their food intake may not be limited to the grain supplies.

Recommendations

Specific field studies, preferably integrated with insect-loss evaluations, should be undertaken to quantify rodent losses in selected environmental situations. Typical sites might be small community or commercial godowns, individual farm storage structures, kitchen or household storage, and field drying or curing operations. Effects of different environmental regimes and different rodent species need to be considered. Whenever possible, association with existing FAO, EPPO, CARE, or binational programs would have obvious advantages.

At the village household level, direct measurements of loss contamination could be made on a daily or short-term basis. This requires measurement of foods purchased or taken from stores and analysis of amounts actually available for later consumption. Rodent populations could be evaluated by estimating sign or intensive removal trapping. Such an effort would require exceedingly good cooperation of village residents and merchants and great honesty on the part of all participants.

For small godowns the most satisfactory measure is the comparison of input stores to those taken out at a later date. This involves measurement of the total stores and evaluation of contamination. For larger godowns, this requires use of sampling techniques. Rat populations have to be determined by trapping or census feeding. Because of the inaccessibility of many areas, use of sign probably would not be satisfactory.

Considerations in evaluative efforts (6, 7) should include: obtaining known, estimated, or "felt" losses from owners, occupants, or merchants; evaluating structure for harborage and infestation potential; quantifying the rodent sign; evaluating daily/weekly/monthly/annual grain-handling procedures and sanitation practices; monitoring incoming and outgoing products to determine depredations; accounting for hoarding activities (eg, burrow excavation); and segregating losses from moisture decrease, insects, birds, and fungi, and determination of primary causes of loss.

Literature Cited

1. JACKSON, W. B. Evaluation of rodent depredations to crops and stored products. EPPO Bull. 7(2): 439 (1977).
2. FAO/CAB. Crop loss assessment methods. FAO Manual, Commonwealth Agr. Bureaux, Slough, England (1971).
3. DAVIS, D. E. The rat population of New York, 1949. Am. J. Hyg. 52(2): 147 (1950).
4. DAVIS, D. E., and FALES, W. T. The rat population of Baltimore, 1949. Am. J. Hyg. 52(2): 143 (1950).
5. TAYLOR, T. A. Major problems affecting productivity of cereals — the pest problem. In: Agr. Res. Priorities for Economic Development in Africa, Abidjan Conf. 1968. NAS-NRC Publ. 2: 175 (1968).
6. ANONYMOUS. Group for Assistance on Storage of Grain in Africa Seminar on the Methodology of Evaluation Grain Storage Losses. Trop. Stored Prod. Inf. 24: 13 (1973).
7. ADAMS, J. M. A guide to the objective and reliable estimation of food losses in small scale farmer storage. Trop. Stored Prod. Inf. 32: 5 (1976).

Bibliography

BROWN, R. Z. Biological factors in rodent control. U.S. Public Health Service Training Guide (1960).

EVERARD, C. O. R. Some aspects of vertebrate damage to cocoa in West Africa. Proc. Conf. on Cocoa Pests W.A.C.R.I. (Nigeria), p. 114 (1964).

FELLOWS, D. P., and SUGIHARA, R. T. Food habits of Norway and Polynesian rats in Hawaiian sugarcane fields. Hawaii. Plant. Rec. 59(6): 67 (1977).

FERNANDO, H. E., KAWAMOTO, N., and PERERA, N. The biology and control of the rice field mole rat of Ceylon *Gunomys gracialis*. FAO Plant Prot. Bull. 15: 32 (1967).

FOOD AND AGRICULTURE ORGANIZATION. Reducing post-harvest food losses in developing countries. ACPP: Misc./21:15 pp + annexes (1975).

FRANTZ, S. G. The behavioral/ecological milieu of godown bandicoot rats — implications for environmental manipulation. All India Rodent Seminar, Siddhpur (1975).

HOPF, H. S., MORELEY, G. E. J., and HUMPHRIES, J. R. O. (eds.). Rodent damage to growing crops and to farm and village storage in tropical and subtropical regions. Centre for Overseas Pest Research, Trop. Prod. Inst. (1976).

KRISHNAMURTHY, K. Problems of food grain storage. Training Manual. Asian Productivity Organization, APO Project TRC/IX/73, Tokyo, 81-84 (1974).

PRAKASH, I. Rodents and their control. Post-harvest prevention of waste and loss of food grains. Training Manual. Asian Productivity Organization, APO Project TRC/IX/73, Tokyo, 185-192 (1974).

SANCHES, F. F. et al. Annual Report. Rodent Research Center, College, Laguna, Philippines (1971).

SPILLETT, J. J. the ecology of the lesser bandicoot rat in Calcutta. Bombay Natural History Society (1968).

CHAPTER VI

D. Rodents

Part 2. Loss Determinations by Population Assessment and Estimation Procedures

J. H. Greaves

Direct measurement of postharvest grain losses to rodents is difficult. As explained in Part 1, the losses to rodents have to be distinguished from losses to birds, spillage, and pilferage, and, in the fields, from shedding or preharvest losses. Therefore, to determine the loss to rodents, all of these other losses must be identified and measured separately. Weight losses due to other pests and to changes in moisture content must also be measured and considered. In addition, specialized studies in the ecology of the rodents may be required. Thus, direct assessment of losses to rodents is complex, and can rarely be contemplated except as an aspect of a multidisciplinary research study.

In contrast, techniques for the estimation of rodent populations, developed by specialists in the fields of rodent control and small mammal ecology, are well established. Clearly, the extent of grain loss to rodents depends on the distribution, size, and species composition of the rodent populations involved. Simple versions of established population assessment techniques can therefore enable the ordinary competent biologist with a little specialized training to derive loss estimates which, though indirect, will be based on objective data and, though approximate, will generally be of the correct order of magnitude.

The methods proposed here are intended primarily for use in grain stores. They may also be considered for use, if intelligently adapted, in fields during the immediate postharvest period and in threshing yards. They are unsuitable for use where the grain, prior to threshing and still attached to the straw or haulm, is either stored in large compact stacks or on vehicles during shipment. The aim of the methods is to estimate the weight of grain consumed by rodents; related losses attributable, for example, to contamination, health hazards, and damage to sacks must be evaluated by other means.

Personnel and Training

The work, including all practical operations such as placement, setting, and checking of traps, should be performed by zoology graduates, preferably with some experience in the fields of rodent control, grain storage, or small mammal ecology. They must possess or first acquire various skills in order to carry out the following operations competently:

a. Identify the rodent species, and distinguish adults of the smaller species from juveniles of the larger species.
b. Identify and evaluate signs of rodent infestation.
c. Set traps.
d. Handle live rodents.
e. Keep field records of the high standard required for investigative work.

These skills are best acquired on the job under the guidance of an experienced specialist. The basics may also be learned in a week or so of laboratory and field training at an institution specializing in rodent control and ecology, in which case it will be necessary to add a further self-training period of 2-4 weeks in which to practice and improve the newly acquired skills in an operational setting.

Selection of Study Sites

The methods given in Appendix B should be employed. Frequently it will be found that appropriate government departments maintain registers of farms, premises of licensed grain traders, etc., which can greatly facilitate selection of a representative sample of study sites.

METHOD A — Preliminary Survey of Infestation

A preliminary survey of the study site must always be made in connection with the two detailed techniques to be described subsequently (METHODS B and C). In addition, a systematic survey of a random sample of sites can, by determining the incidence of sites on which rodents are present and have access to grain, make a valuable contribution to an overall assessment of the rodent problem. It is emphasized, however, that the METHOD A survey procedure will lead to a valid estimate of the quantity of grain lost to rodents only if it is followed up with either METHOD B or C.

Equipment

1. Electric flashlight/torch.
2. Tracking powder (talcum or finely powdered chalk). A glass jar with a perforated lid provides a convenient means of dispensing the powder.
3. Clipboard and record sheets.

Procedure

Two visits will be required. On the first visit record the following information on a record form:
   a. Date of survey
   b. Address of store
   c. Commodities stored and quantities (by weight)
   d. Nominal capacity of the store (by weight)
   e. Date of inward shipment
   f. Expected date of outward shipment
   g. Estimated annual turnover (by weight)
   h. Brief description of the storage structure and conditions of storage
   i. A sketch map of the store (made on the back of a form) showing important features and the location of the stored grain.

Inspect the site thoroughly for signs of rodent infestation, including burrows, excreta, smears, footprints, damage to the commodity or structure, and places where rodents may enter the store. Record these signs on the sketch map as they are found. During the inspection, whether or not signs of infestation are found, lay tracking patches approximately 200 × 300 mm at intervals along the walls of the store and beside the stacked grain, especially around corners. The tracking patches should be laid at the rate of approximately one per 50 tons of grain, except that in stores of less than 250 tons, not less than five patches should be laid. The tracking patches should be entered in a numbered sequence on the record sheet and their positions indicated on the sketch map.

The second visit should be made the next day and the presence or absence of rodent tracks on each tracking patch recorded. Usually it will also be both useful and possible to record whether any tracks found were made by large or small rodents (rats or mice) or by rodents of both sizes. It will not normally be permissible to conclude which species is present until several trapped specimens have been identified.

A simple estimate of the incidence of infestation may be calculated when a random sample of stores of a single type has been surveyed, as follows:

$$\text{Percent of stores infested} = \frac{\text{No. of stores infested}}{\text{No. of stores surveyed}} \times 100$$

$$\text{Percent standard error} = \sqrt{\frac{(\% \text{ stores infested} \times \% \text{ stores not infested})}{\text{No. of stores surveyed}}}$$

METHOD B — Trapping to Extinction

In principle, if a complete census of the population is made by trapping all the rodents that have access to the grain, then the feeding capacity of the population, and hence the current daily grain loss to rodents, can be estimated by multiplying the number of rodents by their daily food requirement, since it may reasonably be assumed that rodents with access to stored grain will use it as their primary food source. The method is suggested for use in stores with populations of up to 200 rodents; this would include a fairly heavily infested store holding up to 500 tons or larger, more lightly infested stores. For larger infestations an alternative technique for population estimation (METHOD C) is advocated.

Equipment

The following equipment is needed in addition to that specified in METHOD A.
1. 200 snap traps (rat size; striking bar 70-80 mm long).
2. 200 snap traps (mouse size; striking bar 40-50 mm long).
3. Spring balance (100 × 1 g).
4. Spring balance (500 × 5 g).
5. Blackboard chalk for marking trap locations.
6. Bait (see later).

Procedure

First make the preliminary survey (METHOD A). The objective is next to trap out the population as rapidly as possible and in a period not exceeding 21 days; to achieve this, the bulk of the population should be caught in the first week. The correct siting of traps is helped by knowledge of the movement patterns of the rodents. Much will be known from the preliminary survey, but it is essential to increase and update this knowledge while trapping is in progress by the temporary placement of extra tracking patches, which should be renewed regularly. The tracking patches will also show, by the absence of tracks, when all of the rodents have been caught.

A large number of traps must be used, at least equal to the supposed size of the rodent population and preferably exceeding it by a factor of 2 or more. They should be distributed at intervals of 1 m or less in all places where the presence of rodents is suspected. Each investigator should be able to deal with about 100 traps daily. Place the traps in a systematic sequence (called the "trap round"), numbering and entering each placement on the record sheet and chalking up the trap number boldly nearby to make it easy to locate on subsequent visits. The bait should be of a sticky consistency such as peanut butter, crushed fruit (banana, oil palm pericarp, or melon), or sweetened dough, and should be pressed firmly into the bait hook so that the rodents cannot simply lift it off but are induced to exert some lateral or downward force on the release mechanism while getting the bait. Succulent baits are often particularly attractive to rodents in the dry environment of a grain store and it may be worth changing the type of bait used after a few days. The traps should be set as finely as possible.

Each day check the trap round and record the species and body weight of each rodent caught for each trap. Every trap, whether it makes a capture or not, must be freshly baited and reset each day and, if judged to be advisable, its position adjusted so as to increase the chance of making a capture. Where both large and small rodents are present, concentrate first on trapping the larger rodents and, as their numbers decrease, gradually switch to using the smaller traps.

It may sometimes happen that though the vast majority of rodents are trapped, a few recalcitrant individuals evade the efforts made to capture them. The size and species composition of this residual population, provided it is very small, can often be estimated from the frequency and size of footprints on the tracking patches. Such estimates and the evidence on which they are based should always be clearly stated.

Grain Loss Assessment

The primary data which should be reported are the numbers and body weights of each species of rodent trapped. The data for each species should be divided into two body-weight classes: 50 g or less, and more than 50 g. The biomass (sum of the body weights) of each weight class should then be obtained for each species. The estimate of the daily grain loss attributable to each species is obtained by multiplying the biomass of the rodents in each weight class by a factor representing the daily grain requirement of a rodent in that weight class, and then adding together the two products.

Preferably the daily grain requirement of each species of rodent in the two weight classes should be determined (as a proportion of body weight) for the commodity and country in question by measuring the actual amounts consumed by representative samples of captive rodents in cages. Where facilities for this are lacking, however, it will generally be adequate to base the calculation on an assumed grain consumption equivalent to 7% of body weight for rodents weighing more than 50 g and 15% of body weight for rodents weighing 50 g or less. The estimated daily grain loss attributable to species "A," for example, would then be (0.07a + 0.15b) g, where a = biomass (g) of rodents of species A weighing more than 50 g, and b = biomass (g) of rodents of species A weighing 50 g or less.

The total estimated daily grain loss is then readily determined by adding together the estimates for the different species, and should be expressed both as an absolute amount and as percentages of the amount of grain in the store and of the nominal capacity of the store. If it can be assumed that the rodent population was reasonably stable, then the loss over a period of time can easily be calculated. Estimates of the annual loss expressed as percentages of the amount of grain actually stored, of the nominal capacity of the store, and of turnover are usually of particular interest.

METHOD C — The Lincoln-Petersen Method of Population Estimation

This method (1) is based on the following principle: First a sample of animals is caught alive, marked, and returned to the original population. When a second sample is then taken, the number of marked animals in the second sample has the same ratio to the total number in the second sample as the number of marked animals originally released has to the total population. Since both the number of marked animals originally released and the proportion of marked animals in the second sample are known, the size of the total population can easily be calculated. The application of this principle to estimating rodent populations involves making several assumptions about the behavior of the populations. In practice the two most important of these assumptions are that 1) the duration of the study is sufficiently short that no significant change occurs in the population, and 2) the chance of capturing a rodent in the second sample is independent of whether or not it is marked. In the typical grain storage situation, the first assumption may be satisfied by completing the study in a period not exceeding 21 days. The second assumption may be satisfied by using live-capture traps for the first sample and snap traps to collect the second sample, since the behavioral responses of rodents to the two types of trap are relatively independent of one another.

Equipment

The following equipment is required in addition to that specified for METHODS A and B.
1. 100 live-capture traps (rat size).
2. 100 live-capture traps (mouse size).
3. Simple restraining devices to hold live rodents for marking (see later).
4. 2 pairs of dissecting scissors.

Two types of live-capture trap are suitable. These are the funnel-type, multiple-catch trap with a horizontal counter-poised door operated by the weight of the rodent as it approaches the holding compartment, and the single-catch trap with a door-closing mechanism operated by a treadle. Live-capture traps actuated by a bait hook are not recommended. Live-capture traps for mice should be made of sheet metal or of 7 mm or finer wire mesh. Specialist advice should be taken if there is any doubt about the suitability of the trap designs available.

Procedure

First complete the preliminary survey (METHOD A). The operation is next carried out in two stages.

Stage 1 lasts 10 days during which the aim should be to capture, mark, and release as many rodents as possible. Distribute, bait, and set the live-capture traps, recording the trap round as in METHOD B. An average density of one rat-sized and one mouse-sized trap per 9 m² is suggested. Fresh bait (eg, soaked grain or fruit) must be provided daily. One investigator should be able to service 50-100 traps. Every morning, each newly caught rodent must be marked by clipping off the middle digit of the right hind foot. To do this, the rodent should be transferred from the trap to a cloth bag where it is restrained gently, while the mouth of the bag is opened to give access to the foot. Alternatively, larger rodents may be restrained in a cylinder or cone made from chicken wire, while mice may be grasped directly with the forefinger and thumb by the loose skin over the neck, either straight from the trap or after first transferring them from the trap to a box or bin 500 mm deep. Newly marked rodents should be released at the point of capture and their numbers and species recorded beside the trap entry on the record sheet. Previously marked rodents should be released at the point of capture without making any additional record.

Stage 2 also lasts 10 days during which the objective is to snap-trap as many rodents as possible, using the procedure described under METHOD B. The body weight, species, and presence or absence of a mark should be recorded for each rodent trapped. In accordance with conditions, a lower trap density may be permissible; however, for the purpose of making satisfactory population estimates it is desirable to recover at least 20 marked rodents of each species in Stage 2.

Population Estimates and Grain Loss Assessment

The primary data which should be reported are:
- The numbers of each species marked in Stage 1.
- The numbers of marked rodents of each species trapped in Stage 2.
- The numbers of unmarked rodents of each species trapped in Stage 2.
- The species and body weight of each rodent trapped in Stage 2.
- The population estimate (P) for each species as $P = an/r$ where $a =$ number marked in Stage 1, $n =$ total number caught in Stage 2, and $r =$ number of marked rodents caught in Stage 2.

The estimate of daily grain consumption is obtained as before, except that it is necessary to determine the weights and relative sizes of the two body-weight classes by reference to the sample of rodents trapped in Stage 2. Thus, where in the absence of data from captive rodents it is assumed that the daily grain consumption figures for animals greater than 50 g and for smaller rodents are respectively 7 and 15% of body weight, the daily grain loss attributable to species A will be:

$$P [0.07ab + 0.15 (1-a) c] g$$

where P = the population estimate for species A,
     a = the proportion of rodents of species A of body weight greater than 50 g,
     b = the mean body weight (g) of rodents of species A weighing more than 50 g, and
     c = the mean body weight (g) of rodents of species A weighing 50 g or less.

(The parameters a, b, and c must be calculated from the sample trapped in Stage 2.)

If the population estimate, P, is unsatisfactory owing to fewer than 20 marked rodents of the species concerned having been trapped in Stage 2, then the data for two or more species may be pooled to give a combined estimate. The estimate of total daily grain loss should be expressed in the various ways suggested under METHOD B.

Literature Cited

1. LE CREN, E. D. A note on the history of mark-capture population estimates. J. Anim. Ecol. 34: 453 (1965).

CHAPTER VI

E. Measurement of Losses Caused by Birds[8]

This section recognizes that there is scarcely any line between grain held in the field for maturing and drying and grain held for maturing, drying, and storage. The storage portion of the cycle is intertwined with both the drying and holding requirements. At times grain, chiefly maize, sorghum, and millet, may be held for extended periods in the field prior to harvesting for storage or direct to the table use. Some of the most serious grain losses occur at this stage when losses to *Quelea* spp., parakeets, and blackbirds have assumed disastrous proportions; however, losses are rarely quantified.

It is often difficult to relate specific birds to designated damage or losses. Feeding patterns may be irregular or overlap; insect outbreaks, drought, or flood may alter expected patterns; fungi may enter as a secondary factor related to bird damage; and the measurement techniques, themselves, may be tedious and exacting. Comparisons of damage to benefits, whether off-season removal of weed seeds compensates for food losses, effects of intensive monoculture, the mutually destructive breakage or cutting of heads by mammals and birds, and other matters all complicate loss assessments. Losses to piled and bagged grain are often observed but rarely if ever quantified, and birds are usually more readily accepted than rodents as part of the environment.

While losses are real, satisfactory methods of determining losses have seldom been available or used. The most intensive statistical efforts have been on blackbird damage in the United States. These have used the detailed row-centimeter measurement technique and visual-loss estimates as summarized below:

Row-Centimeter Measurements (used on maize). The number of damaged and undamaged ears in a row (15-100 ft) are counted. On damaged ears, the average lengths of damaged and undamaged kernel rows are measured to the nearest, approximately, 2 or 3 mm. These lengths are converted to losses per area, eg, tons/hectare. Less exacting are simple measurements of the portion of ear damaged, which may require some arbitrary averaging if the damage pattern is not symmetrical.

Visual-Loss Estimates. This technique is usable on many different crops, but observers must be trained and their procedures calibrated for each crop. This is a much more rapid technique, since counting is not specifically required. Damage-level criteria (5, 10, 20, 40%) are established and workers trained by repeated tests to distinguish visually between these levels of damage/loss.

Losses to stored bagged or bulk grain can best be measured by before and after weights over a period of time. The kinds and numbers of birds and how much time they spend on the grain should be noted. These figures can then be used in estimating losses in similar situations elsewhere.

---

[8]This brief summary was excerpted and added to by K. L. Harris from Estimates of Bird Depredations to Agricultural Crops and Stored Products by W. B. Jackson and S. S. Jackson, first presented at the Colloquium on Crop Protection Against Starlings, Pigeons, and Sparrows, European and Mediterranean Plant Protection Organization, Jouy-en-Josas, France, Oct. 18-20, 1977.

CHAPTER VI

F. Moisture Measurement

T. A. Granovsky, G. Martin, and J. L. Multon

Accurate measurement of grain moisture and its variations is critical for proper assessment of weight losses during storage. Changes in moisture content are accompanied by changes in weight and volume and need to be recognized as separate from actual grain losses. Frequently, the weight of moisture gained or lost by grain may exceed weight losses induced by insects, rodents, birds, or fungi. Moisture changes are merely the gain or loss of water; the others may alter food quantities or qualities. Therefore, measuring the moisture content of grain is an extremely important operation from three standpoints:

1) *Technology:* Knowledge of moisture content is needed to efficiently determine and manage the harvesting, drying, stocking, and processing operations. It is also essential for assessing and controlling postharvest losses insofar as the action of water governs deterioration phenomena.

2) *Analysis:* To compare the results of analysis with a fixed basis (dry matter or standard moisture content). In particular, assessing the weight of a stock of grain and making loss determinations requires accurate knowledge of the moisture content.

3) *Marketing:* Commercial purchasing and sales contracts often stipulate an upper limit for the moisture content not to be exceeded.

Samples should be analyzed as soon after being obtained as is practical. Since grain can gain or lose moisture rapidly, all samples not immediately tested should be retained in air- and moisture-tight containers and not exposed to undue temperature variations.

It is necessary to emphasize how important it is for all measurements to be made with thoroughly standardized procedures. The International Association for Cereal Chemistry has called attention to various procedures for measuring moisture content.

Moisture measurements depend on two fundamental baseline procedures. These procedures determine what water shall be classified as free moisture in the grain and, hence, is the water that is dealt with in a percentage moisture determination and will be the basis of the reading given by a moisture meter. Table V summarizes international approval of the two types of baseline methods.

An in-depth discussion of the comparative values of the "fundamental reference methods" versus the "practical reference methods" is not within the scope of this manual. They do, however, involve highly specialized apparatus and conditions (see ICC standards in Table V).

Use of Meters (See also Appendix C)

The amount of grain necessary for determining moisture content will depend on the testing method used. Some methods are portable and enable determinations in the field. Other methods are laboratory-based and may require a constant power supply and chemical agents. Selection of a meter will

TABLE V

**Present State of International Standardization of Reference
Methods for Measuring Moisture Content of Cereals**

	Fundamental Reference Method Vacuum Drying, P_2O_5, 50°C		Practical Reference Method Drying at 130°C	
			2 hr	4 hr
	Soft Wheat, Durum Wheat, Oats, Barley, Rye, Rice, Maize, Millet, Sorghum		Soft Wheat, Durum Wheat, Oats, Barley, Rye, Rice, Millet, Sorghum	Maize
International Association for Cereal Chemistry[a]	Standard 109-1 (July 1978)		Standard 110-1 (July 1978)	
International Standards Organization	ISO-R-711 (April 1968)	in progress	ISO-R-712 (April 1968)	in progress
European Economic Community			ICC-110 adopted regulation 130/67 EEC	in progress
International Legal Metrology Organization			International recommendation No. 8 (Oct. 68)	in progress
International Seed Testing Association			Provision No. 9 Proc. ISTA 31 (1966) No. 3	

[a]See Multon, J. L. et al. (1978) Répétabilité de la méthode de référénce pratique de dosage de l'eau. Revue de Metrologie Pratique et Legale, Nantes, and Guilbot, A. et al. (1973) La détermination de la teneur en eau des semences. Seed Sci. Technol. 1: 587. A copy of the ICC Standards is available on request from the ICC Secretary General's Office, A-2320 Schwechat, Schmidgasse 3-7, Vienna, Austria.

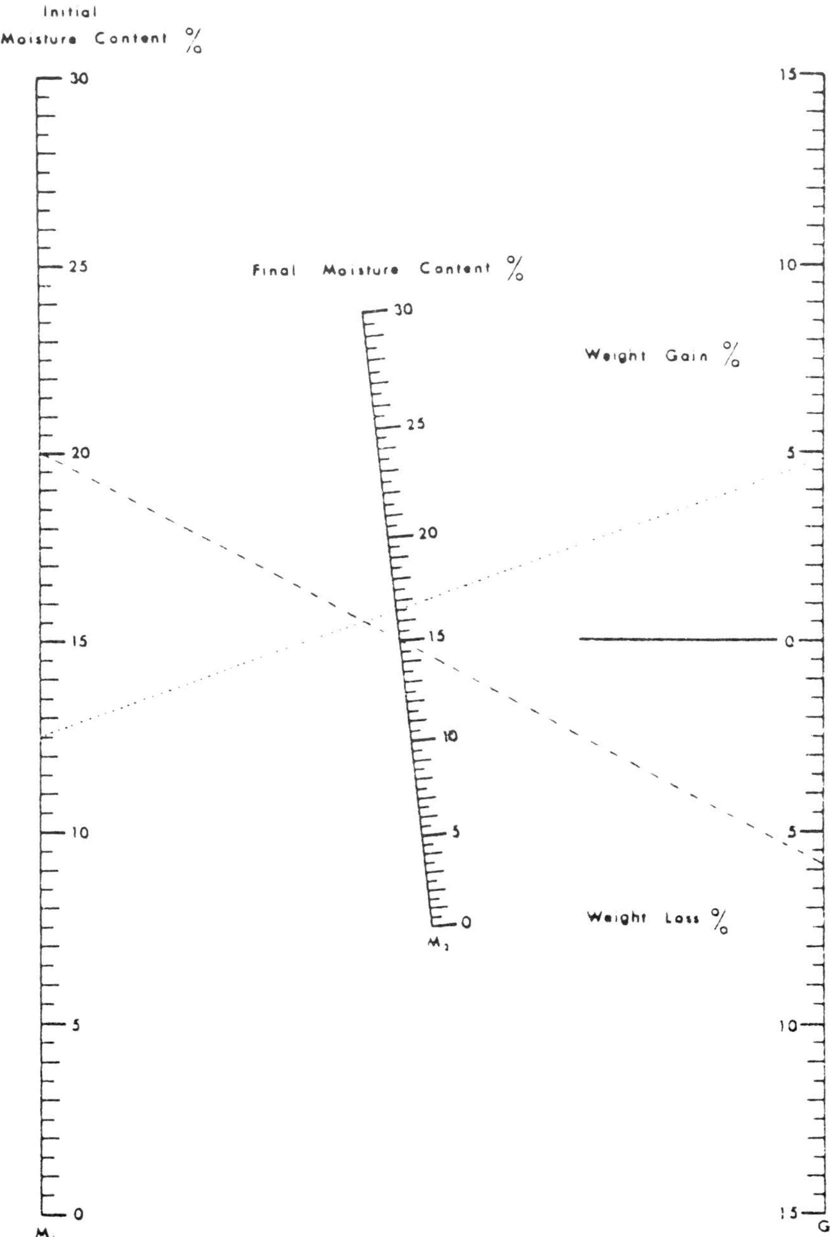

Fig. 10. Nomograph for calculation of grain weight changes resulting from changes in grain moisture content. (From M.F.S. Jamieson. Trop. Stored Prod. Inf. 20: 19 [1970].)

depend on where the determinations are to be made. In general, use of a moisture meter is encouraged, especially one which is both portable, enabling on-the-spot moisture determinations, and rugged enough to withstand transport from locale to locale. Aspects of proper adjustments and sensitivities of each meter should also be considered when making a selection. The data in Appendix C are pertinent in deciding what meter an investigator will select, amount of grain needed, speed of operation, and accuracy of each. In any case, manufacturer's directions in using the meter should be followed.

Moisture meters require periodic calibration, the frequency of which will depend on the meter and conditions of its use. Often it can be checked against samples especially prepared and packaged for this purpose. In other cases, it may be taken to a central laboratory for comparisons with a meter reserved for this purpose, for comparison with control samples, or for comparisons with results by standard oven-dry methods. To re-set, manufacturer's directions should be followed.

As a general rule, all field or laboratory determinations should include at least three and preferably five replicates in an effort toward greater validity. Consistency in handling and preparation of samples for moisture content determinations is indispensable.

The percentage loss or gain in weight by the grain may be derived from the average initial and average final moisture contents. The nomograph (Fig. 10) is employed as follows:

1. Lay a straight edge so that the initial and final moisture content values lie along this edge.

2. Read the percentage gain or loss in weight off the right-hand bar.

For instance, if the initial moisture content of a sample is 12.5% and the final moisture content value obtained is 16.5%, then the percentage weight gain is about 4.8% (represented by dots). Conversely, if the initial value is 20% and the final value only 15%, then a 5.9% weight loss has been realized (represented by dashes).

VII. OPERATIONS STANDARDIZATION AND CONTROL

A. Handling of Samples in the Laboratory

T. A. Granovsky

When a sample arrives in the laboratory from the field, it should be in a sealed moisture-proof container and at ambient laboratory temperature when opened. This will require proper preparation and care of samples in transport, field to laboratory, in addition to prompt attention upon arrival by the laboratory staff.

During handling in the laboratory, each sample *must retain its identity* as to location, data collected in the field, grain type, variety, and time in storage at *all times.*

As each sample enters the laboratory, it should be handled as per the sample flow and by the procedures indicated below:

Sample Flow

If Moisture Content was Determined in the Field

1. Sample enters laboratory.
2. Sample collection data recorded on laboratory data sheet.
3. Whole sample weighed (grain, dust, insects, dockage).
4. Grain sieved: insects are recovered and placed in 70% alcohol; dust is weighed, if necessary, and discarded.
5. Weight-to-volume vessel properly filled and weighed.
6. Grain from weight-to-volume vessel recombined with rest of original field sample, repeated five times, and averaged.
7. Sample (1 kg) is divided into a series of 8-32 subsamples.
8. Five subsamples are randomly selected for tests on losses induced by insects as per other instructions (see Section B, Chap. VI).
9. Other subsamples may be used as needed in tests on losses induced by microorganisms/respiration, aflatoxin, etc.
10. All data derived during loss analysis should be recorded on the data record sheet (Fig. 11).

If Moisture Content is to be Determined in the Laboratory

1. Steps 1 to 5 are as above, but after weight of the weight-to-volume vessel has been recorded, the moisture content is determined before the sample is

recombined with the rest of the original field sample. Weighings and moisure content determinations are repeated three times each and averaged separately. Steps 7 to 9 are then finished.

2. The total weight of grain, dust, insects, and dockage should be determined for the whole sample as it arrives from the field. This figure, and all

---

EXAMPLE OF SAMPLE FIELD/LABORATORY DATA SHEET FOR MAIZE

Collection Date_______________________Control Number______________________
                                                    (imprinted number)
Collector's Name____

Farmer's Name____

Name of Location____

Town/City_______________________________County__________________________

Seed Purchasing Date____________________Planting Date___________________

Variety_________________________________Doubling Date___________________

Estimation, by the Farmer, of the area planted in corn___________________

<u>FIELD</u> <u>PORTION</u>

1. Drawing of corn field:

2. By random sampling in the field, gather the 4 data items below:

    n = Total number of rows in the field_________________________________.

    $\bar{x}_1$ =   Av length of a row, in meters$\dfrac{(\quad)(\quad)(\quad)(\quad)(\quad)}{5}$ = __________.

    $\bar{x}_p$ =   Av number of plants/
          meter ever having ears $\dfrac{(\quad)(\quad)(\quad)(\quad)(\quad)}{5}$ = __________________.

    $\bar{x}_s$ =   Av space between rows,
                      in meters $\dfrac{(\quad)(\quad)(\quad)(\quad)(\quad)}{5}$ = __________________.

3. Sampling of ears:

    a. Use a table of random numbers to select 25 plants to be sampled for ears.

    b. Prenumber 25 plastic bags (1 bag/ear) with the numbers obtained from
        the table of random numbers.

    c. Collect at least four sets of 25 ears of corn/field (separate data sheets
        are needed for each set).

    d. Each ear should be placed into the appropriate prenumbered plastic bag
        (1 bag/ear).

    e. Each group of 25 ears from a single row is then bagged together for direct
        transport to the laboratory to collect the data indicated below.

(Figure 11 continued)

<u>LABORATORY</u> <u>PORTION</u>

1. Data sheet for information to be collected from each ear.

Ear Number[a]	No. of kernels completely damaged				No. of kernels partially damaged			Wet wt of Grain[b]
	Insects	Fungi	Birds	Rats	Insects	Birds	Rats	
TOTALS								

[a] If ear selected (based on table of random numbers) is missing, leave blank all spaces to the right of ear number. Try to determine if each missing ear is the result of rodents, birds, dogs, or man. Record observations.

[b] Total weight of grain, insects, dust, etc. from each ear.

2. Total Wet Weight of grain collected (from lower **right-hand box above**)= _________ Wet wt

3. Combine the grain from all the ears, mix grain 3 times in a Boerner or Gamet Divider.

4. Fill the weight/volume vessel, then take the grain weight (g) and the percent moisture content (% mc) of the grain in the sample. Record data below:

	1	2	3	4	5	Av
Weight/Volume (g)						
Moisture content (%)						

Recombine samples and mix well.

Repeat above process 5 times and average, as indicated at right.

(Figure 11 continued)

5. Recombine samples and mix at least 3 times in a Boerner or Gamet Divider.

6. Divide down sample to obtain a series of subsamples, approx. 100 g.

7. Take weights of 100 undamaged kernels from 5 randomly selected subsamples and record data below:

	1	2	3	4	5	Av 100 kernel wt

Weight (g)

8. Special notes on the samples:
For instance, number of insects (immatures and adults) per sample, species of insects present, problems in handling sample (lost or broken samples), damaged container, or spoiled sample.

EXAMPLES OF CALCULATIONS BASED ON THE FIELD AND LABORATORY DATA GATHERED ABOVE

1. Average Wet Weight/ear = $\dfrac{\text{Total wet wt}}{\text{No. of ears collected}} = \dfrac{(\quad)}{(\quad)} = \underline{\qquad}$.

2. Total Dry Weight = (Total wet wt) − [(Total wet wt) (Av % mc)]

   Total dry wt = ( ) − [( )( )] = $\underline{\qquad}$.

3. Average Dry Weight/ear = $\dfrac{\text{Total dry wt}}{\text{No. ears collected}} = \dfrac{(\quad)}{(\quad)} = \underline{\qquad}$.

4. <u>Estimated</u> wet weight corn yield/field = $(n)(\bar{x}_1)(\bar{x}_p)(\overline{\text{xWW}}/e)$

   field data No. 1 above

   $(\quad)(\quad)(\quad)(\quad) = \underline{\qquad}$.

5. <u>Estimated</u> dry weight corn yield/field = $(n)(\bar{x}_1 (\bar{x}_p)(\overline{\text{xDW}}/e)$

   field data No. 3 above

   $(\quad)(\quad)(\quad)(\quad) = \underline{\qquad}$.

6. For other calculations see chapters VI and VIII.

Fig. 11. Example of sample field/laboratory data sheet for maize.

subsequent data, should be recorded on separate data record sheets for each sample. A suggested partial sample sheet for data derived in the laboratory is presented in Fig. 11.

3. The grain is then sieved to separate off insects and dust (depending on the characteristics of the debris, use No. 10 or No. 25 sieve and solid botton pan). Insects should be placed into bottles containing 70% alcohol, labeled as to origin, and identified as required.

4. The weight-to-volume vessel, Fig. 8, should be properly loaded, filled, sliced, and weighed. This is repeated five times and a mean is taken. After each weighing, the grain should be recombined with the original field sample and remixed before another sample is removed (see Section B, Chap. VI).

5. The 1-kg sample is divided into 8-32 subsamples by using a recognized method such as a sample divider or by coning and quartering (Appendix A). It is suggested that the subsamples be placed into individual pre-marked containers to facilitate their manipulation. As noted in Appendix A, subsamples may vary somewhat in size (number and weight of kernels) depending on the commodity and the conditions under which the grain was produced.

6. Five subsamples (replicates) are then selected at random for subsequent tests on losses induced by insects. See Section B, Chap. VI, for measurements of losses caused by insects.

7. Other samples may be used as needed in tests on losses induced by microorganisms/respiration, aflatoxin, etc.

8. All data derived during loss analysis induced by insects, microorganisms/respiration, rodents, birds, and physical losses should be recorded on the data record sheet (Fig. 11).

CHAPTER VII

B. Operations Manuals and Laboratory Records

T. A. Granovsky and K. L. Harris

In the conduct of any survey there is absolute need for an operations manual that describes how the survey is to be managed to ensure that the purposes of the project will be performed. Operations manuals can be in any useful format, but should specify duties of each employee and operation. Such a manual is designed for internal use by operating personnel.

Depending on the complexity of the operation, the manual may be divided into subsections for on-the-spot use in specific operations. If an operation is large enough to involve a payroll, there should be a division under corresponding functional headings for the purchase of supplies, travel, field observations and sampling, laboratory analyses, and reporting and tabulating results.

A complete compendium of what to include in an operations manual is beyond the scope of this work; however, guidance for a field and laboratory operations manual is given below:

It is imperative that all procedures for information-gathering, sample collection and transport, sample examination and reporting, and collection and tabulation of results be tested in dry runs before the actual information gathering gets under way. This period of preparation is used to give a final assessment of the quality of the written directions, on training or the need for additional training, and on the suitability of individual people, procedures, and forms for the job. Make changes as required.

Field Controls

1. Once the sample collection and field observation sites, system, and criteria are established, these same parameters need to be recorded on paper in terms suitable for the user.

2. Sample collection should be explicitly set forth as to where, when, and how — with no room for deviation.

3. Use of alternative procedures, when permitted or applicable, reporting of inability to sample or make observations, reporting of broken containers, lost samples, and miscounts all need to be explicitly detailed.

4. There need to be observation reporting forms, sample collection forms and labels, packaging and shipping forms, and supplies where required.

5. Triers (see Appendix A) and other technical devices and supplies need to be provided (bags, preservatives, clasps) and their use completely described (see below).

6. Where, when, how, and how much sample is taken needs to be explicitly set forth. How to operate triers, how much preservative is to be added, how to get samples to the laboratory, and speed and route of sample shipments must be established, set down on paper, and controlled.

7. Use of moisture meters, scales, or balances and any special devices needs to be explained stepwise in complete detail, as well as their care and maintenance and checking for malfunction.

8. All reports of all observations and collections are to be on pre-numbered forms or in numbered-page notebooks furnished by the project. All entries should be original and in ink or ballpoint pen with no erasures or data-recording on other slips of paper. All pre-numbered pages and forms must be accounted for with no forms discarded.

9. All entries are to be made directly into the notebook as each measurement is made. Supervisors should check on this immediately upon arriving for a surveillance visit.

Field Observation Form

Form number _________________________
                          (Imprinted number)

Date of observation _______________

Name of site ____

Location ____

Description: ____
        (Depending upon what is pertinent, this is highly variable.

____
It relates to the roof, sides, foundation, etc., or to hillside, valley,

____
village, city, etc.)

Special features: Insects _____________________________

                      Rodents _____________________________

                      Birds _______________________________

                      Sprouting grain _____________________

                      Protection from rain ________________

                      In bags, bins, on the ground, etc. _________

                      ____

                      Amount of grain on hand _____________

                      Etc.

Inspector/observer ___________________ Signature ________________
                 (Print name)

Date submitted _______________________

Fig. 12. Example of a field observation form.

10. Any confusion, or lost or broken sample, should be reported to the immediate supervisor without fear of reprisal or penalty.

11. Suggested data record forms are presented in the following figures: Fig. 12, a sample field observation form; Fig. 13, a sample collection form; and Fig. 11, a sample field/laboratory data sheet for maize.

12. Supervisory field controls require careful monitoring by several varied techniques, such as scheduled and unscheduled supervisory visits, discussions

```
                          Sample Collection Form


                                      Sample number _______________________
                                                      (imprinted number)
                                      Date collected _______________________


Where collected _________________________________________________________


Observation at collection site (insects, rodents, birds; who gave permission to
take sample:

      ___________________________________________________________________

      ___________________________________________________________________

      ___________________________________________________________________

      ___________________________________________________________________

      ___________________________________________________________________

Payment for sample:     Currency and amount _____________________________

                        In kind and amount _______________________________ .

To  whom payment made ___________________________________________________

Where sample held _______________________________________________________

Date delivered ___________________________

To whom _________________________________________  Signature _____________
                  (print name)

Remarks:_________________________________________________________________

      ___________________________________________________________________

      ___________________________________________________________________


Signature _______________________________

Date _____________________________________
```

Fig. 13. Example of a sample collection form.

with various employees and subjects being investigated, and comparisons with automobile logs and daily expense logs and diaries. These techniques can be part of the supervisory operations manual and can be kept as checklists.

Laboratory Controls

1. Once the sample analysis procedures are established, they need to be recorded on paper in terms suitable for the user.

2. Analytical techniques must be followed *to the letter*. No alternative procedures are permitted unless expressly authorized in the operations manual.

3. All needed equipment must be provided and maintained in working order using a recorded maintenance and calibration record.

4. All reports of all tests are to be on numbered forms or in numbered-page notebooks furnished by the project. All entries should be original and in ink or ballpoint pen with no erasures or data-recording on other slips of paper. All pre-numbered pages and forms must be accounted for with no forms discarded.

5. All entries are to be made on-the-spot as the results are obtained. Supervisors should monitor this very carefully.

6. Any confusion, mistake, mixup, lost sample, damaged container or sample, or spoiled sample should be reported to the immediate supervisor without fear of reprisal or penalty.

7. Figure 11 is a sample reporting form.

8. Analytical controls require careful monitoring by several varied techniques, such as scheduled and unscheduled supervisory visits, generally observing operations if the analyses are being done close to headquarters, and comparisons with daily logs and diaries. A supervisor should know the analytical procedures. By watching the operator, the supervisor will form a dependable judgment as to the analyst's expertise and working habits.

9. Analytical operations require the use of internal controls, such as seeded, or pre-set, standardized control samples sent through the analytical procedures with or without the analyst's knowledge, duplicate samples analyzed at different times by different analysts, and supervisors who can check or repeat grain separations and other analyses.

10. All instruments require regular calibration, especially moisture meters and balances for grain loss work.

   (a) Moisture meters usually can be calibrated against a standardized meter in a national or international institute. Standardized held-in-glass or otherwise sealed samples may be obtained from well-known institutions for use in calibration. For periodic use this is more practical than using oven-drying moisture determinations.

   (b) Balances and scales need to be checked against a special set of weights of known value. Frequency of checking depends on accuracy requirements and the usage to which the balance is subjected.

Reporting Results

1. All results should be on numbered forms or in numbered-page bound notebooks.

2. Results should be submitted on a regular basis, and should be checked and otherwise followed as the work proceeds. To allow them to accumulate for an end-of-project or delayed review is to lose an opportunity to find and contain sources of error.

3. Decisions on interim reports and keeping the staff informed of the data need to be resolved on an individual project and person basis. In some cases, being aware of what is happening and working toward overall goals will maintain and improve work equality, although it could introduce bias.

4. Standard terminology of weight loss should be followed. This manual recommends:

$$\frac{ow - cw}{ow} \times 100 = \% \text{ loss}$$

where ow = original weight on a dry weight basis,
     cw = current or final weight on a dry weight basis.

Other formulas, such as those in Chapter VI, Section B, where direct differences cannot be calculated, may have to be substituted.

VIII. APPLICATION AND INTERPRETATION OF RESULTS

A. The Chronological Approach:
Losses as Reflected by Use Patterns

J. M. Adams

In making grain loss estimations, it is important to relate losses to the pattern of grain consumption. If grain is left untouched throughout the storage period, the total loss over the season can be obtained by accurately weighing all the grain in and out of the store and comparing the totals. This does not, however, indicate the relation between loss and time, ie, when the loss reached a peak or whether it was related to a particular part of the season. If at the time of removal the estimated loss is 10%, then this represents the total loss over the storage period. In most cases, however, grain is removed at intervals during the storage period and each quantity removed will have been exposed to deterioration for a different length of time and will have suffered a different degree of loss.

If a measurement of the quantity removed is available, then estimates from samples covering the removal period and pattern may serve to cross check with the total loss as well as showing the pattern of loss.

If, as often happens on subsistence farms, the amount removed is quoted in volume terms (eg, tins), then the volume removed will be the same whether or not the grain is damaged but the weight will be different. In this case, the weight of grain that occupies the farmer's measure should be recorded carefully at the beginning of the storage period. For each subsequent removal of grain, this weight can be reduced by the percentage of loss estimated from the appropriate sample. If samples are taken at monthly intervals and the dates of removals are known, an approximation can be made by applying the estimated loss to removals two weeks either side of the sampling date. To obtain the total loss, all individual losses can be summed.

Where removals are roughly estimated, the loss may be obtained by calculating the percentage of the total quantity stored which was removed at each sampling date and applying the percentage loss to this. The resulting losses are then summed to produce an overall percentage loss, as in Table VI.

When stored grain is regularly removed for household use, weight loss may be measured by taking, or having the user set aside, a sample from, or taken at the same time as, the portion withdrawn for use. The household may be provided with an equivalent amount of grain in exchange for the test samples.

135

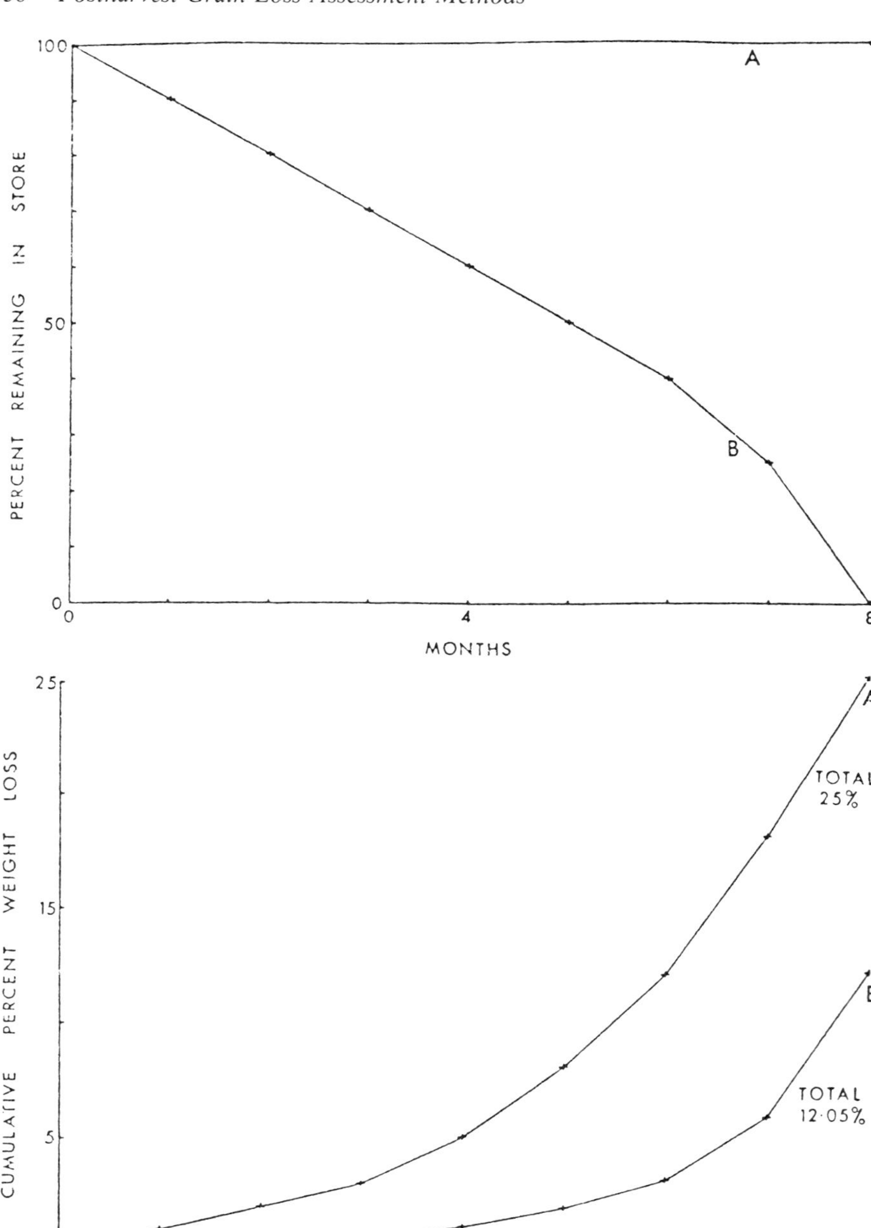

Fig. 14. Effect of rate of consumption by farmer on cumulative weight loss of stored produce. Top: A = no consumption by farmer during storage period; B = pattern of consumption shown in Table VI (when all grain is used over eight months). Bottom: Result of consumption pattern (above) on cumulative % weight loss over eight months. A = no consumption; B = consumption as in B above.

TABLE VI

Relation Between Weight Loss and Consumption

	Months in Store							
	1	**2**	**3**	**4**	**5**	**6**	**7**	**8**
Quantity removed, %	10	10	10	10	10	10	15	25
Weight loss in sample, %	1	2	3	5	8	12	18	25
Weight loss as % of total stored	0.1	0.2	0.3	0.5	0.8	1.2	2.7	6.25
Cumulative weight loss as % of total	0.1	0.3	0.6	1.1	1.9	3.1	5.8	12.05

This is an actual use-weighted loss of 12.05% compared with a loss of 25% (as measured in month 8 of Table VI) if only a single, final visit had been made and there was no allowance made for consumption (see Fig. 14). Line A of the top of Fig. 14 represents a farmer who holds a quantity of grain in store for sale when the price is high and does not remove any until the date of sale, when the store is completely emptied. Line B represents a subsistence farmer who regularly removed grain from the store for family consumption. The total loss in weight suffered in case B is considerably reduced because a decreasing proportion of his total stored grain is exposed as the level of loss increases with time.

The same procedure may be adopted in relation to nutritional loss, bearing in mind that damage may cause greater losses in preparation of food where soaking of the grain is involved. It may also be used to evaluate quality loss in terms of money. For seed grain, the loss is the drop in germination from the time of storage to the date the seed is required and is simply the difference between the percentage germination recorded on the two dates.

CHAPTER VIII

B. Losses and the Economist

M. Greeley and G. W. Harman

Definition

To the economist, storage losses refer to changes in the value of grain which occur as a result of any physical change while it is in store. Alterations involving biological changes normally reduce its value and thus involve an economic cost. Losses may also occur during marketing and, to the extent that waste and unintended physical alterations take place, during the primary processing of grain.

Setting Terms of Reference

The economist evaluates loss by assessing the cost or sacrifice borne as a result of its occurrence. Since losses can occur at various points in the marketing pipeline and will, if significant, have consequences for individual stores-owners and consumers, merchants, marketing boards, etc., and to the country as a whole, it is essential to define from whose viewpoint the assessment is to be made. In this guide, concentration is centered on the consequences of losses for stores-owners at farm level in developing countries.

An attempt should be made to approximate the magnitude of the value of losses before time is spent on trying to reduce them. If this value proves to be low, expenditure of appreciable resources on reducing losses may not be justified. Even when it is established that losses are sizable, consideration should be given to the relative desirability of their reduction compared to alternative investments. If the purpose is to increase the quantity and quality of grain available to users, there may be other more practical and cost effective ways of achieving this end. Examples of possible alternatives to improving storage are measures to stimulate the use of fertilizer to increase production of grain and changes in the marketing system to encourage stores-owners to store less by making grain/flour available for their purpose at a fixed price throughout the storage season. On the other hand, there are situations where no alternatives are available, and food lost equals people starved. These situations are difficult to resolve on an economic basis.

In assessing the practicability of improving storage, it is essential to have the improvement tested by stores-owners since this may reveal unanticipated problems. One important aspect of this testing will be to determine if storage owners are sufficiently motivated to undertake improvements in their storage methods. Factors affecting acceptability and utility of storage improvements are not all predictably quantifiable and require practical testing before their benefits can be accurately assessed. National priorities may meet individual needs and vice versa. Distribution of potential benefits should also be taken into account since these will vary appreciably according to the type of improvement proposed and the point in the marketing pipeline at which it is made.

Nature of Losses

The physical alteration and diminution of grain in store will affect its weight or quality. Both changes will alter its value and should be assessed separately. Change to the nutritional worth of grain may be regarded as a particular type of quality loss. Such losses are only relevant to the owner if they affect the price of grain that is sold or are of sufficient size to reduce the value of grain in other ways. An example is a reduction in an owner's capacity to work which may occur through eating grain which has suffered a nutritional loss.

Losses may involve other economic costs by necessitating expenditure to reduce them and by affecting the timing and, therefore, the price of grain that is sold. Major factors influencing the economic consequences of loss are: scarcity of grain, the extent of seasonal price fluctuations, the time at which the impact of loss is felt, the proportion of a crop that is stored, the extent of the premium on better quality grain, and the opportunities for using damaged grain in other ways.

Collection of Data

The objectives of data collection are to ascertain, by examining the behavior of procurers, handlers, and stores-owners, the consequences of losses incurred by them; and, if the level of losses justifies changes in the system of storage, to assess the likely costs and benefits of such changes.

The exact nature and amount of information to be collected will depend on the circumstances in each situation and the time at the researcher's disposal. The basic minimum data necessary for reliable evaluation are as follows:

1. Use of stored grain, preferably throughout the whole storage season
   - amount consumed by stores-owner and his dependents
   - amount sold; price obtained
   - amount used for other purposes such as for seed, feedingstuffs, making beer, payment of wages

In obtaining these data, attention should be given to the reasons for usage at a particular time, interrelations between usage and type of store, effects of a stores-owner having more/less grain available, influence of the variety/type of grain on its use, and the effect of varying degrees of physical damage on usage. If regular visits are not being made to an owner, data will need to be collected on the time at which grain from the store was exhausted and the consequences of this for the owner. In examining usage, grain sales to any marketing authority should be distinguished from those at village level since the price received will probably be different.

In addition to the amounts of grain used in various ways, the pattern of usage should also be noted in relation to other stores which the owner may possess. For example, is grain taken out of one store until empty or is it taken out of more than one? Why?

2. The marketing system
   - method of operation
   - factors determining prices received, particularly timing of sale and quality of grain (including any statutory regulations applicable)
   - influence of variety/type of grain

3. Behavior of stores-owners
   - motivation for growing grain
   - degree of knowledge of losses
   - measures (if any) taken to eliminate loss
   - capability and motivation for adopting any suggested improvements in storage
   - work undertaken off farm; its nature, timing, and remuneration, ie, is the store the major source of food grains?
4. Stores/storage practices (existing and as would occur when suggested improvements are included)
   - materials used in construction, quantity and price
   - time taken to collect materials and build or improve the store
   - season at which constructed and alternative work at that time (on and off farm)
   - expected life of improved or traditional store
   - insecticides used, quantity and price
5. General
   - purchases of grain, reason, timing, amounts, prices
   - type/variety of grain grown/stored
   - cost of seed

Methods of Collecting Data
   - published reports and economic data
   - discussions with those having detailed knowledge of the behavior and practices of stores-owners
   - questionnaire surveys.

Training of field staff who will be conducting, or assisting in conducting, questionnaire surveys should receive close attention to ensure that they thoroughly understand the questions to be asked and the reasons for them.[9] If at all possible, all field staff should be accompanied on initial visits, and periodically thereafter, to ensure that questions are put without bias. Questionnaires should be tested experimentally on a sample of participants before a complete survey is made so that misunderstood questions can be rephrased or removed.

Use of Data in Evaluating Losses

1. Weight Loss

The value is obtained by pricing the weight loss according to the use to which the lost grain would have been put and the effect of its loss on the stores-owner. For example, if the grain would have been consumed by the owner, its replacement cost as food would normally be used; similarly, if sold, its sale price, and, if used for seed, its cost of replacement.

---

[9]This matter was debated at Slough and no real consensus obtained. Some felt that information gatherers should *not* understand their questions and that the most reliable information was obtained when information was gathered in a fixed mechanical manner.

2. Quality Loss

This may be assessed by adopting a standard of quality and measuring loss as the difference between this standard and that of the grain in the store. The relevant standard will depend on the intended use of the grain but often it will be that set by a marketing authority. If no such authority exists, an attempt must be made to examine how the grain usage is affected (if at all) by the existence of differing qualities. The standard which affects its usage should then be adopted.

The economic cost of the quality loss will be represented by:

$$Lq = V_s - V_a$$

where Lq = value of quality loss,
V_s = value of grain if it was all of a standard set,
V_a = value of the quality of the grain in store when used.

Quality loss of grain intended to be used as seed is especially serious. If the stores-owner does not realize that it is damaged, it may be planted and result in a lower rate of germination. This loss is assessed as the difference between the value of the crop expected from the undamaged seed and that which would be produced from the damaged seed.

3. Indirect Loss

This is the cost of any insecticide or other treatment used by the stores-owner to minimize his losses.

4. Nutritional Loss

This can be valued in the same way as quality loss by adoption of a standard. Since this method is liable to a high degree of subjectivity, the reasons for using a particular standard need to be clearly stated. In some cases, nutritional loss will not reduce the economic value of grain to an owner; for example, it may not, taken by itself, necessarily reduce its sale price.

5. Other Losses

Stores-owners may suffer other economic costs due to losses, but the valuation of these will be specific to particular circumstances and it is not possible to provide more than the general principles of valuation already outlined.

Further Points to Note on Evaluation

1. Valuation should be based on the time when the impact of loss is felt by the owner. This will not necessarily be at the time when the loss occurs. This factor will be of particular importance in cases when the price of grain fluctuates appreciably during a storage season.

2. In arriving at a final loss figure, the value of damaged grain in any alternative or secondary use should be considered. For example, if grain in-

tended for human consumption was damaged and, therefore, used to feed cattle, the loss suffered by the stores-owner would be:

$$Ln = Lf - Lc$$

where Ln = net loss,
    Lf = value as food,
    Lc = value as feedingstuff.

Summation of the different types of economic costs occurring as a result of physical loss will provide an estimate of the total economic impact of losses. Such estimates should be related to the "wealth" of the stores-owners concerned since losses of the same value will affect poorer stores-owners to a greater extent. In this respect, care should be taken in quoting average values.

Use of Data in Assessing Improved Method(s) of Storage

The benefits of a system of storage are assessed by a comparison of the costs involved with its output as measured by a valuation of grain leaving the store. Improved storage may be reflected by a reduction both in weight and in quality losses per unit of storage cost. The value of any additional amounts of grain made available by a reduction in weight loss should be based on the use to which this extra grain would be put. The value of the reduction in quality losses is obtained by grading grain stored in both the normal and improved manner as it leaves the store using a common standard. The amount of qualitative benefit will be:

$$Qb = Vi - Vu$$

where Qb = qualitative benefit,
    Vi = total value of grain leaving improved store,
    Vu = total value of grain leaving unimproved store.

In assessing the reduction both in weight and in quality losses, it is necessary to ascertain the level of these before improvements in storage are made. Care should be taken that the figures obtained are representative since there may exist appreciable variation between different seasons and stores. The costs involved in adopting a particular system of storage may be divided into those of materials and labor used in constructing the store and of any treatments applied to the grain. The cost of any purchased inputs, including labor, will be the actual amount paid. Any time spent by the stores-owner or his family on constructing the store or treating the grain should be priced at a theoretical or imputed wage rate. The rate used will normally reflect the wage being offered in a type of occupation similar to that in which the stores-owner is engaged. This rate should be taken only as a general guideline. The objective in using any particular one is to express the cost (if any) to the owner of the time which he and his family spends on storage by the value of the time given up on its alternative use. In some cases, materials used to build a store will not be purchased but gathered from fields or woods. The cost of these free goods in evaluation should be that of the time spent in obtaining them.

In assessing the cost of time, attention should be given to the seasonal pattern of agricultural activity and also to the fact that the value of time at a particular period may differ between different stores-owners according to the amount of land and labor at their disposal.

The three main methods of relating costs of benefits are by means of a ratio (cost-benefit ratio), a rate of return, or by comparing the additional benefits from taking a particular action with the additional costs incurred. The last of these approaches is particularly suitable where the changes to an existing system of storage are relatively small. The rate of return concept is more suited to situations in which changes to the system of storage is extensive and sizable capital investments are involved. Where the rate of return concept is used, the value of grain removed from a store will be expressed as a percentage of the store's cost. Finally, but importantly, if benefits gained over a period of years are being compared with costs incurred at a point in time, they must be discounted using a suitable rate of interest. The spread of benefits within the total period is a significant factor in this procedure.

CHAPTER VIII

C. Conversion into Monetary Values

E. Reusse

After having been physically and quantitatively assessed, food losses have to be expressed in monetary terms. This is necessary to establish a common denominator for cost-benefit analysis, in which cost (investment in potential improvement measures) and benefits (expected reduction of food losses) can be weighed against one another.

Thus, if a farmer can reduce his storage loss from 8 to 4% by means of fumigation, and the fumigant plus amortization of plastic sheetings amount to $3 per 500 g, then 1 kg of grain must be worth more than 15 cents to warrant the investment. If a rice miller can raise the extraction rate of paddy rice from 63 to 66% by additional installations (including rubber rollers) and additional controls by qualified technicians, together increasing milling cost from $2 to $2.50 per 100 kg, then 1 kg of rice must be worth more than 17 cents to make the improvement financially feasible. While the financial value of the rice to the miller might be only 15 cents per kg, the economic value for the national economy of the country concerned may be much higher, as it is when the rice gained through the advanced milling technique can serve the substitution of imports, thereby freeing valuable foreign exchange.

The question is how to determine the value of a unit weight of grain. The financial value can be one value for the individual innovator (the farmer, trader, or processor, whether private, cooperative, or state enterprise) and a differing economic value for the economy as a whole. The viewpoint from the individual enterprise sphere is also referred to as the micro-economic consideration, as opposed to the macro-economic one taken from the viewpoint of the national economy.

Food losses occur principally at three different levels: farm, wholesale and processing, and retail. These levels are linked by transport. The gains in time-, form-, and place-utility added to the food product at and between the various levels, carrying those essential inputs as storage, transportation, processing, packaging, financing, risk-bearing, and logistics decisions, add value to it. The cumulative value added in the postharvest system for storable food crops in developing countries generally amounts to between 50 and 100% of production cost, depending on distribution radius and degree of processing involved.

In a competitive marketing system, the value added is reflected in the market price received for the food product at the various levels of the process. A typical postharvest cost-price structure for rice might be as shown on the chart on the following page.

It follows that the physical loss of 1 kg of rice in the form of paddy occurring at the farm level in financial terms represents only 57% of the loss of the same quantity of rice, after milling, at urban retail level. It is, therefore, vital to value a food loss at the farm gate or market price prevailing for that stage of processing and for that geographical area where it occurs. For transport-inflicted losses, the market price at point of destination would apply; for milling losses, the price for the milled product would apply.

	Cost per kg paddy	Cost per kg milled rice (at 66% ext. rate)
farm gate value	10	
+ transport	1	
rural assembly market value	11	
+ bagging, transportation, etc.	1.5	
provincial market value	12.5	
+ milling cost	1.5	
milled rice (in terms of paddy)	(14.0)	21
+ bagging, transportation, etc.	(1.6)	2.4
urban wholesale market value	(15.6)	23.4
+ packing and other retail cost	(1.8)	2.7
urban retail market value	(17.4)	26.1
urban retail market price	. . .	27

Since market and farm gate prices are subject to seasonal fluctuations, when working at a national level, annual average prices should be used. To eliminate abnormal annual crop situations, the average over the past three years may best be taken. An inflation factor, however, should be added, if necessary, since implementation of any remedial measures will usually be delayed.

So far we have discussed the financial valuation of food losses typical for micro-economic consideration; let us now look at a few major situations where under macro-economic consideration the financial, price-based valuation has to be corrected or substituted by an economic valuation. As the examples will show, these situations typically arise because of government intervention in the price structure:

1. The situation of subsidized producer or consumer prices
    a. Subsidized producer (farm gate) prices: for economic valuation the subsidy element has to be eliminated (*downward* correction of financial values).
    b. Subsidized consumer prices: same applies, but *upward* correction of financial values.

2. Overstated official foreign exchange rate of national currency: In such situation domestic price development lacks close correlation to world market prices. This fact has little relevance in a closed food economy, ie, where the country is neither a regular exporter nor importer of the staple food crops (products) in question. In an open food economy, however, where food losses are affecting the foreign exchange intensive marginal area of export surplus or import substitution, those losses have to be valued at the average annual FOB export or CIF import price, respectively, under application of a shadow rate of the foreign exchange involved in converting to national currency values, shadow rate being understood as the rate expected to prevail under conditions of free floating exchange rates. The FOB or CIF value thus established in national currency has to be deflated by the transportation cost between the geographical area where the field losses are occurring and the seaport. This would include the simplifying assumption that, in most developing countries, consumption of imported staple foods is concentrated in geographical areas near ports of importation.

3. A shifting area between financial and economic valuation is entered when food losses in government reserve stock and price stabilization schemes have to be valued. Since the selling price of those stocks in most cases is related neither to market value nor internal cost price calculation, the value applied to stock losses should at minimum reflect the full unit cost of operation, including accumulated storage cost over the recycling period which may extend over two to three years.

APPENDIX A

SAMPLING GRAIN

1. Comments on Probing Techniques and Probes

a. In this volume the terms trier, probe, thief, and spear are used interchangeably.

b. A compartmented grain trier should be used that will reach the bottom of the container with each compartment 15 cm long (see Fig. 15). Noncompartmented grain triers should not be used to sample grain.

c. In probe-sampling a bin from the top, the probe or trier should be inserted in the grain at an angle of about 10 degrees from the vertical, with the slots closed. The probe should be opened while the slots are facing upward. While the slots remain open, the probe should be moved up and down so that all openings may be filled. The probings should be emptied onto a sheet and coned or quartered or mechanically divided to sample size.

d. In bag sampling, the trier should be inserted from a corner diagonally across to the farthest corner.

2. Techniques for Sampling Bagged Produce[10]

P. Golob

The Importance of Sampling

Quality is an important factor which dictates the value of a commodity. It is judged by the overall appearance of the produce and will be adversely affected if there are holes in the grains caused by insect attack, discolored grains from mold damage, shrivelled grains, cracked and broken grains from bad handling, or rodent hairs and droppings.

Infestation by stored product insect pests before harvest is common so that a consignment may enter a store having a low-level infestation. Depending on climatic conditions, the pests can multiply rapidly and greatly damage the crop. Thus it is of vital importance that the infestation be detected as early as possible, preferably before storage begins. The crop must be inspected and sampled as it is unloaded from lorries or railway trucks before it is stacked for storage.

As a commodity deteriorates during storage, it loses value. For the government of an exporting country this can mean a loss of foreign exchange. For the subsistence farmer the losses result in less food to eat. Poor storage conditions can aid the increase of insect and mold populations and bad storage structures can allow the entry of rodents. It is, therefore, important to continually check stored produce in order to monitor changes.

For practical reasons it is not physically possible to examine every grain in a consignment. Thus the quality of the whole has to be judged on the basis of a sample. The sample must be representative of the individual bag or stack from which it is drawn. In this Appendix the various techniques which may be used

---

[10]Adapted from Trop. Stored Prod. Inf. 31: 37 (1976).

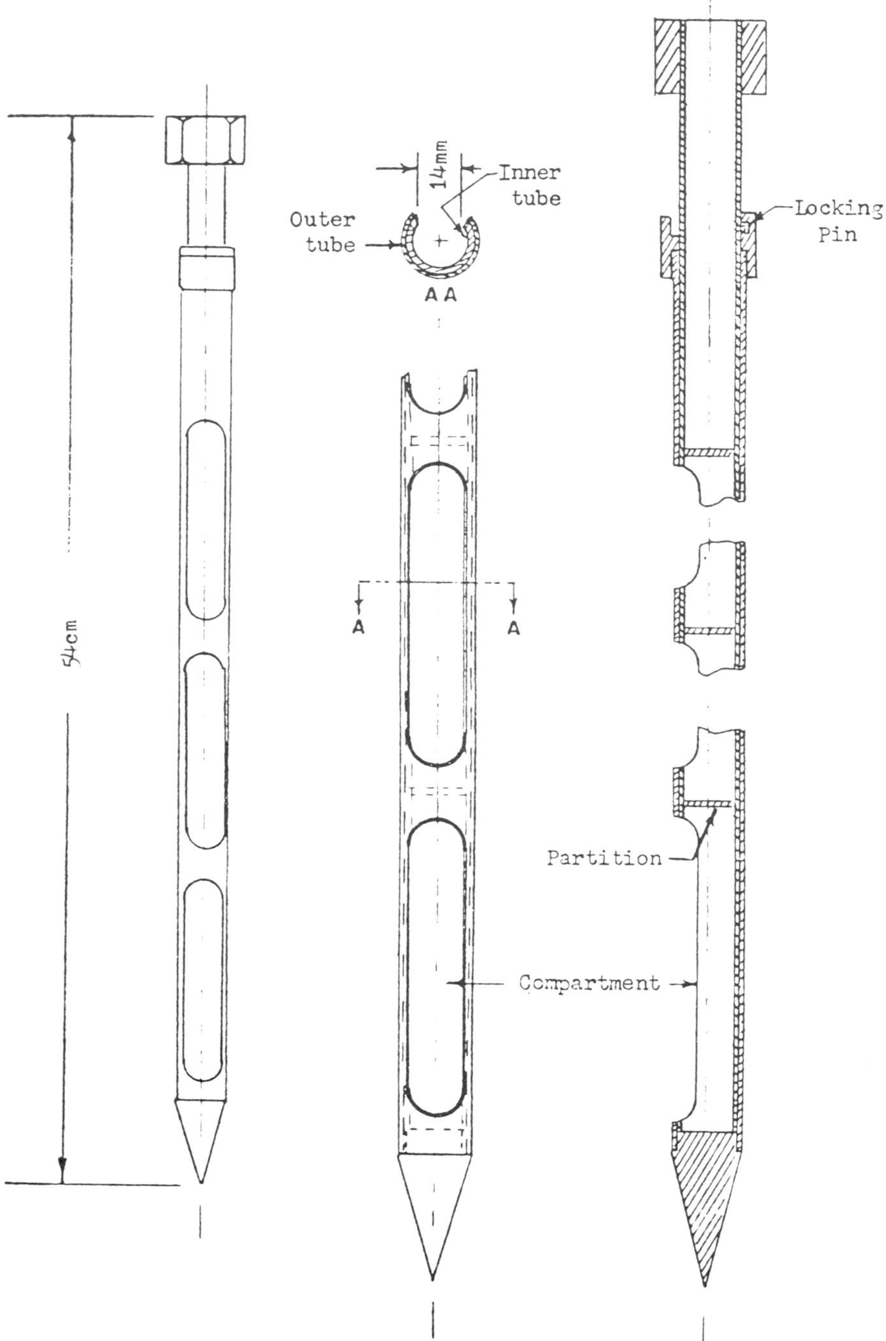

Fig. 15. Compartmented bag probe. (From U.S. Department of Agriculture GR Instruction 916-6, Exhibit F.)

to obtain representative samples from bagged commodities are described and their limitations are discussed.

Sampling From Stacks (See also Chapter IV)

The principles of sampling from stacks apply to all types of stacking situations whether in a large warehouse or godown, a ship, a train or lorry, in a trader's store, or a farmer's crib. In practice, however, it may not be possible to put all the principles to use due to the accessibility of the stack.

Consignments of produce can be divided into sectors on the basis of location. For example, in a ship the commodities may be segregated in different holds, wherein each hold can be regarded as an individual sector in terms of climatic and other physical influences. Similarly each boxcar of a train might be regarded as a single sampling entity. Each sector must be identified and sampled individually.

As the conditions within each sector may fluctuate as much as those affecting the total consignment, it is important to obtain samples which are representative of the sector from which they have been drawn. Each sector itself can be stratified and samples must be drawn from all areas within each sector, ie, from the top, middle and bottom, left and right, center and periphery. Removing samples from these strata should be performed at random.

Twenty-four sampling points from a cuboidal stack should provide an accurate representation of the stack. However, taking samples from as many points as this for all but the largest stacks is uneconomical and unwarranted, as fewer sampling points will give as accurate a pattern. Five sampling points are recommended for wagons and lorries of up to 15 tonnes, eight points for up to 30 tonnes, and eleven points for containers up to 50 tonnes, as shown in Fig. 16.

Number of Sacks From Sector (See also Chapter IV)

The above recommendations are inappropriate for sampling stacked bags because they regard the stack as a two-dimensional structure. They take no account of the difference between top layers in a container and lower layers where dust and insects would tend to accumulate. They disregard any possible changes that affect one side of the stack rather than the other and they ignore the fact that stacks are accumulations of individual units that can be separately sampled.

Practical experience has shown that the optimum number of samples to be obtained from a large consignment (over 100) of sacks is given by the square

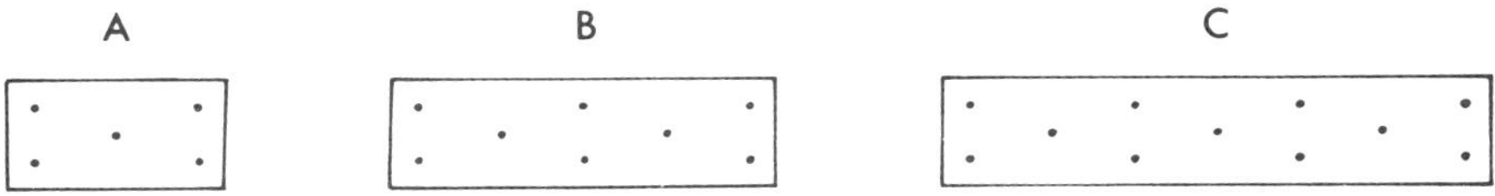

Fig. 16. Sampling points in a container as recommended by the International Organization for Standardization (1969, 1972). A = Five sampling points for wagons or lorries up to 15 tonnes; B = eight sampling points for wagons from 15 to 30 tonnes; C = 11 sampling points for wagons from 30 to 50 tonnes.

root of the total. Jelier (1) suggests that for sectors of 10-100 bags, 10 bags should be taken at random and for up to 10 bags, each bag should be sampled. Thus, from a lorry having perhaps 100 sacks the sample would consist of 10, which would represent all areas of the stack.

Bags drawn from the stack using the above rules constitute the initial sample which should be taken randomly but at the same time should be representative of the whole stack. In practice, when obtaining the initial sample from a small stack as found on a lorry, it is not possible to sample entirely at random. The structure and size of the stack determine from which areas bags must be chosen, so that the number of bags from which the random choice has to be made is relatively restricted and may only be three or four bags.

In many cases it is not possible to sample randomly from all sectors of very large stacks. Only by breaking the stack down would most bags become available. Thus only a relatively small area can be sampled. Effort should be made to get to bags in the middle. To do this, several layers of bags at the top of the stack should be removed and a bag in the sixth or seventh layer obtained for observation. This practice in no way utilizes randomized searching for initial samples.

The Bag as the Sample Unit

Sampling of the commodity within the bag must be random so that every grain has a chance of being picked. Many of the sample-taking procedures are not random but tend to be haphazard, resulting in having human bias. With haphazard sampling, such as using a spear or trier, every grain does not have a

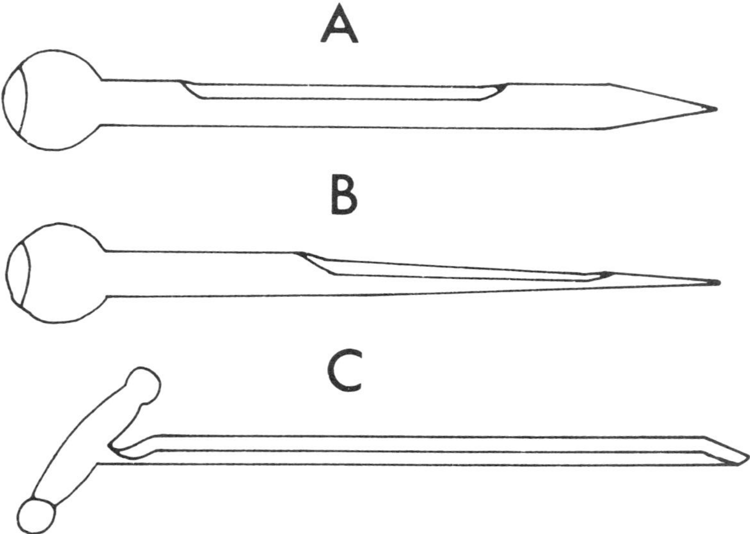

Fig. 17. Typical spears for sampling bagged grain. A = Closed spear for sampling large particles, such as maize or coffee; B = closed spear for sampling small particles, such as wheat or rice; C = open spear.

chance of being picked as just a portion and not the whole bag is the sample unit.

Methods of Obtaining Samples From Sacks

1. Spear Sampling

Bag sampling with a spear or trier is practiced throughout the world. There are many types and variations of sampling spears, the commonest of which is illustrated in Fig. 17. Bag spears are usually cylindrical in shape and between 40 and 45 cm in length with a diameter of 2.5 cm, except at one end which is drawn to a point.

The tube is open on one side to allow grains to fall into a collecting channel, which passes back along the length of the spear and opens out through the handle. This type of spear is used for collecting large particled material, such as maize grains or coffee berries. Other types of spears may be of similar design but narrower for collecting smaller grains such as wheat and sorghum (Fig. 17B) or simply open-grooved lengths of metal attached to a handle (Fig. 17C).

The spear has several good features; it is cheap, simple to use, and is a quick way of obtaining grain from bagged produce. The tip of the spear is pushed into the bag, and the body with the open side face down is inserted for the required distance. Grain is collected in the channel by twisting the spear so that the open side is turned upwards. On withdrawing the spear from the bag, the grain is tipped out of it into a container. If the spear is inserted into the sack at an angle, with the point uppermost, grain entering it can pass straight into a container without the spear being removed, so that a large sample can be obtained (see Fig. 18). Generally six or more samples are removed from each sack to make up a primary sample.

Because of its widespread use, the faults of spear sampling are usually disregarded. However, the disadvantage of the spear is so fundamental that it

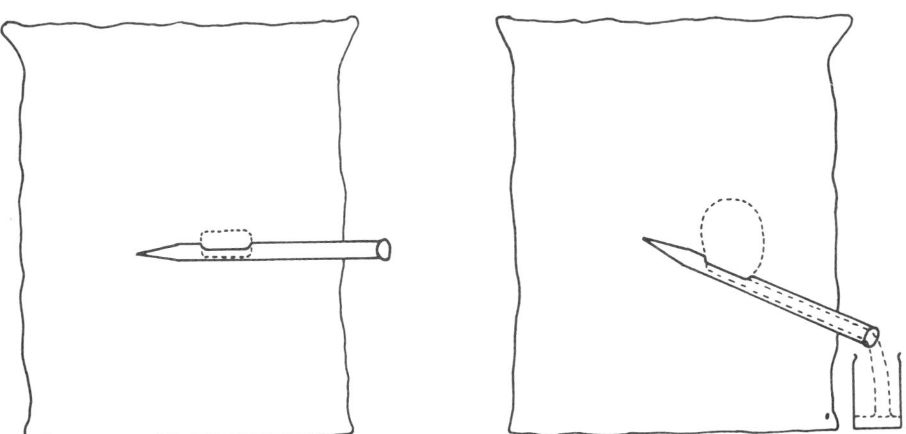

Fig. 18. Area of produce sampled when a spear is introduced into the sack.

negates most of the results obtained upon analyzing samples collected by this method. When a spear is inserted into a sack either horizontally or at an acute angle, only a very small volume of the sack commodity is sampled, ie, precisely that material that actually falls into the spear cavity. The sack is not sampled randomly; the grains picked depend on the haphazard method used to insert the spear into the bag (see Fig 19).

Many elements of stored crops (such as protein and vitamin contents) are generally constant throughout a single sackful of produce or any variations that do occur are insignificant. Produce moisture content and insect numbers, however, may not be constant throughout the bag. Insects, in particular, distribute themselves neither uniformly nor randomly. They are often found in pockets associated with the dust or meal material at the bottom of the bag or in areas of local heating and wetting.

Producing information on insect numbers in a sack using a spear sample can lead to erroneous conclusions and be totally misleading, either overestimating a population or more frequently underestimating it. Examples of the way in which this could occur are shown in Fig. 19. In Fig. 19A, a large population of insects crawling on the bottom of the bag could easily be missed by spear sampling; it is difficult to sample very close to the bag fabric, top and bottom. Observing or missing a population such as this could influence the decision to treat the commodity to eradicate the infestation, resulting in heavy losses of produce. In Fig. 19B, small pockets of two or three insects could by chance be picked up by a spear sample. Six insects in 100 kg of maize may not require eradicating if the produce is not going to be stored for long periods. However, six insects in a 500-kg sample is equivalent to 1,200 individuals in a 100-kg bag if randomly distributed, whereas there may be less than ten in the whole bag.

Thus spear sampling can produce grossly misleading results and should be

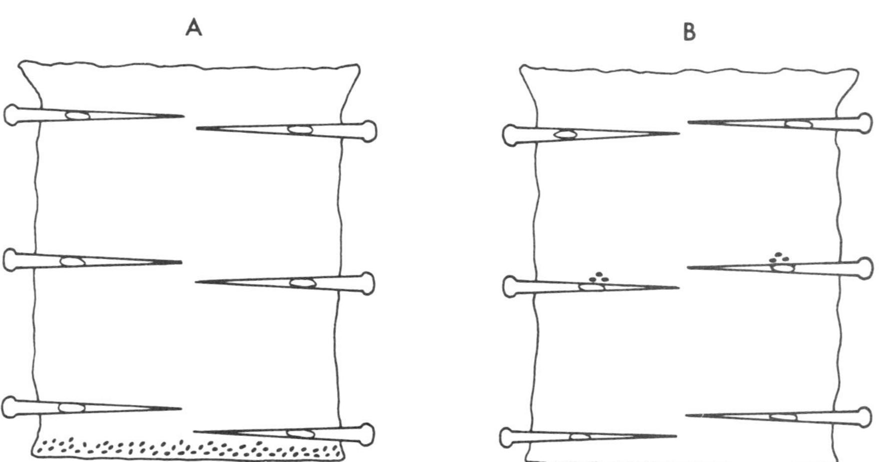

Fig. 19. Inadequacies of spear sampling (black dots represent individual insects). A = Large populations can be underestimated; B = small populations can be over-emphasized.

avoided. A compartmented probe (Fig. 15) should be used whenever a probe sample is taken. Compartmented probes are available in bag size, as in Fig. 15, or in larger sizes for probing deeper piles in bins, wagons, etc.

2.   Coning and Quartering

Sampling at farmer and trader level requires a procedure that is simple, cheap, and accurate. Coning and quartering is such a method.

When a bag of commodity is opened and the produce is tipped onto the floor, the contents naturally assume the shape of a cone. By shovelling material from the periphery of the cone to the apex, while circling the periphery, complete mixing and randomization of the produce will occur. This mixing needs to be done for 3 to 4 min at least five times round the circumference. Division of the bulk into halves and then quarters using a flat piece of wood or quartering irons produces four samples of very similar properties. From a 100-kg bag, each sample would be 25 kg, too large to be useful. By further subdivision, using the same coning and dividing procedure, each quarter can be divided into 1/8th, 1/16th, 1/32nd, etc., subsamples.

Sampling error by coning and quartering is about 10%, which is much more accurate than spear sampling. This method is time-consuming, however, and can only be used when a small number of bags require sampling. For continuous sampling at marketing board or export level, the produce flow sampler can be used.

3. Sieving

The three techniques described above comprise methods by which small quantities of material can be removed from the bulk for analyses or inspection. An estimate of dust content or insect number in a sack can best be obtained by using a sieve. Unlike the methods discussed above, a sample representing the whole sack is not obtained. Instead the commodity is divided on the basis of particle size. Smaller particles, including insects, pass through the sieve mesh whereas large particles pass over it and are returned to its bag.

A type of bag sieve is shown in Fig. 20. The produce is tipped into a hopper located above the sieve mesh. On oscillating the mesh by a simple hand-cranked gear mechanism, the produce flows out of the hopper and over the mesh surface. The bulk of the produce passes back into the sack, and dust and insects are collected in a tray slung below the mesh. The mesh size can be altered as required depending on the particle size of the produce being sieved. Tests have shown that more than 90% of all dust and insects is removed using this apparatus, the recovery of insects being independent of the population density.

Apparatus for Sample Reduction

Sample reduction can be performed by coning and quartering (see above) or by using specific apparatus designed for this purpose. Generally, this equipment divides the sample into halves which then have to be passed repeatedly through the divider until a workable sample is obtained. Such a divider is the Boerner divider (Fig. 21).

1. Boerner Divider (Conical Type)

This is a gravity mechanical divider which works on the same principles as the produce flow sampler (see below). The produce flows out of a hopper and around a cone but, unlike the PFS which takes a single sample of up to 12% of the total, the Boerner simply divides the total in half. Instead of the four sampling points of the PFS, the Boerner has a series of channels around the periphery of the cone. As the commodity flows into the channels, it is directed into one of two collecting points. The direction of flow of the channels alternates around the periphery so that every other one directs the flow into the same collecting pot. The Boerner is an accurate method of sample division.

2. Box Divider

A simplified version of the alternate-channel separation is the box divider shown in Fig. 22. It is less expensive than the Boerner, more easily transported, less subject to damage (and when damaged more easily repaired), and does

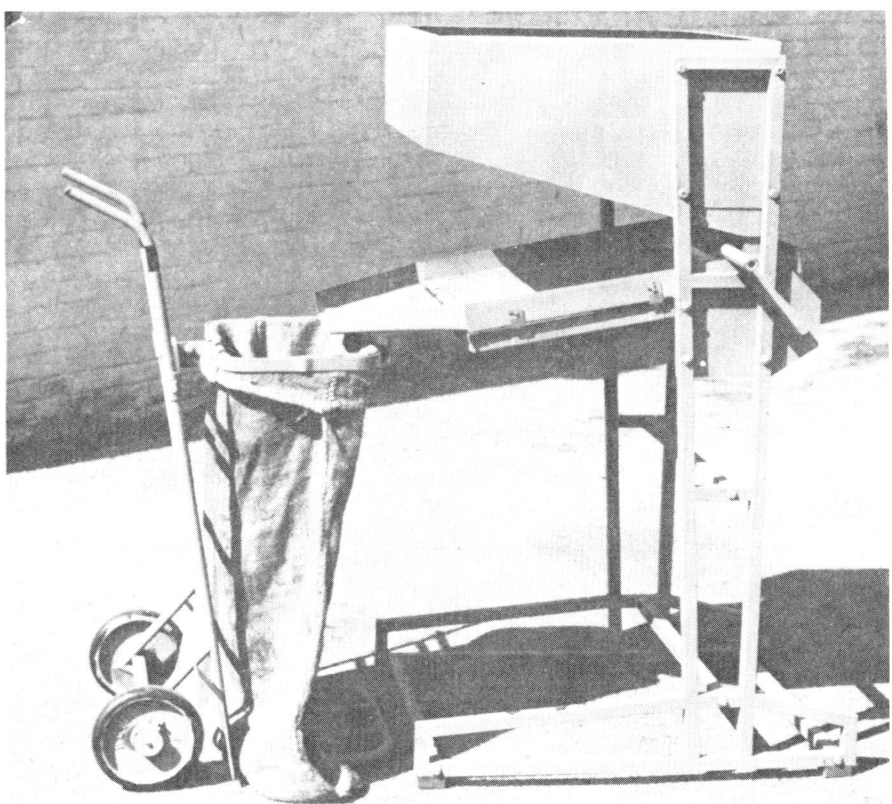

Fig. 20. Sack sieve.

almost as accurate a job as the Boerner. In using, care should be taken that the slot widths remain uniform and are not bent out of position.

3. *Motorized Divider* (Centrifugal Type)

In this divider the seed falls into a shallow rotating cup from which it is flung into a chamber divided into two or more outlets at the bottom. An example of this type is the Gamet divider (Fig. 23). In dividers of similar design, the grain may be delivered from a rotating spout over a number of

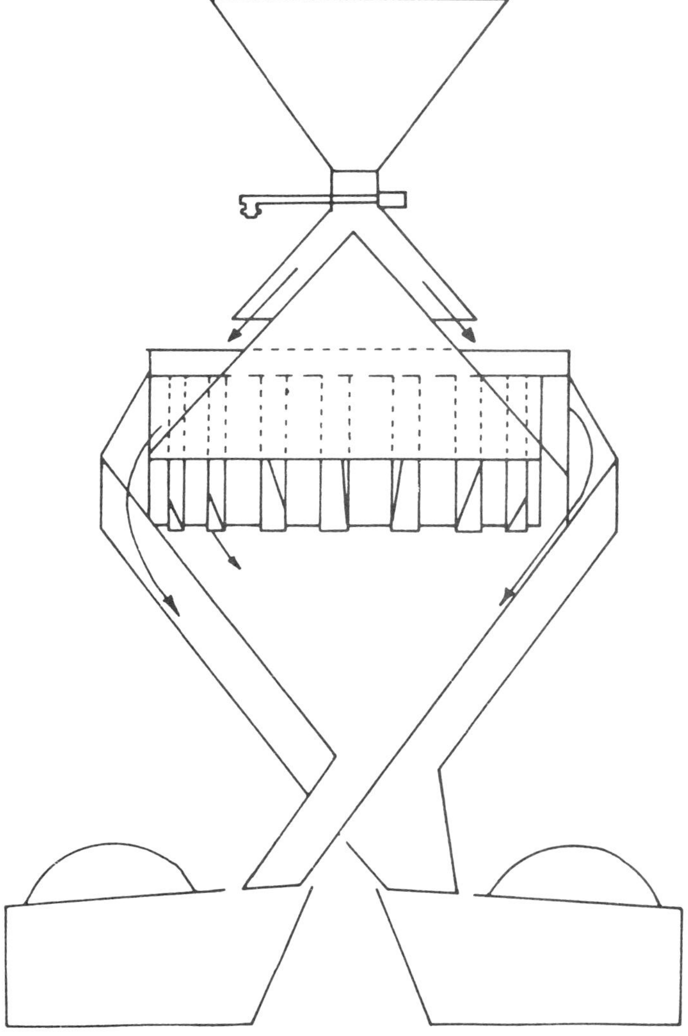

Fig. 21. Boerner divider.

containers or over a cone with adjustable dividing blades at the bottom which may be arranged to separate off any desired fraction.

4. Produce Flow Sampler

The produce flow sampler (PFS) (Fig. 24) is a device designed by the British Tropical Stored Products Centre for taking samples from whole bags of grain. The produce is tipped into an upper hopper which has an opening at the bottom. The opening is closed by a bung until sampling commences. On removing the bung, the produce flows down and around a cone and, because the apex of the cone is placed exactly under the center of the hopper opening, the flow of produce is equal all around the cone. Samples are separated from the main flow at four points at the base of the cone, the points being spaced equally around its periphery. The bulk of the produce is recollected in a sack attached by hooks to the bottom of the collecting funnel. Sampling time is 20 sec for a 100-kg bag. Size of the samples can be altered by changing the vent that covers each sampling point.

The PFS was originally designed for sampling bags as they were off-loaded from lorries before the produce went into store. For this purpose, the PFS is 8

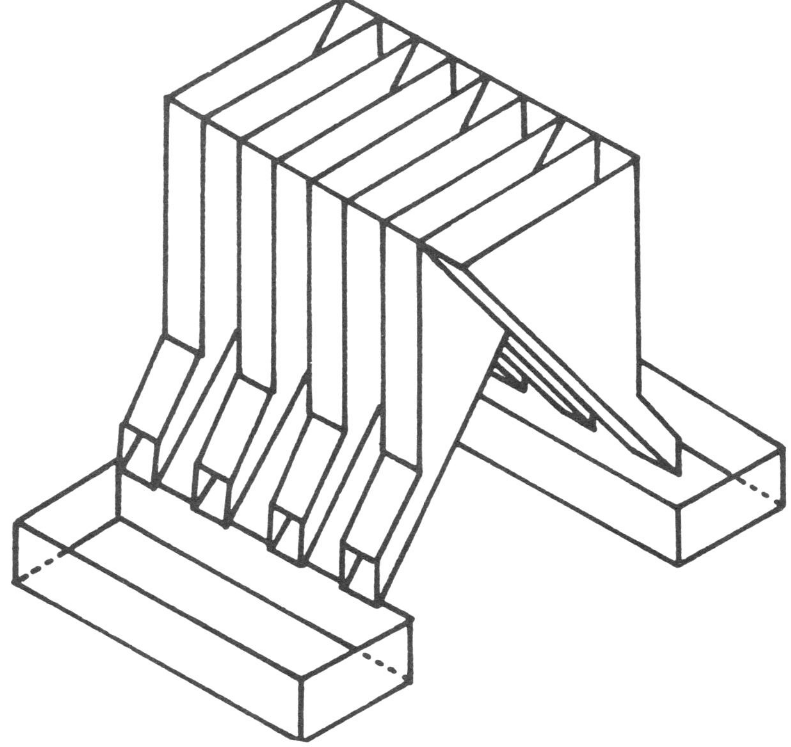

Fig. 22. Box divider.

ft high but the length of the legs can be lowered if required. All flowable commodities can be sampled using the device.

Tests on the accuracy of this method have been performed using bags of produce containing a small percentage of grains stained with a dye strategically placed at different parts of the bag to simulate pockets of defective produce. With groundnuts, for example, containing 5% dyed kernels, the percentage of stained nuts in the samples ranged between 3.4 and 6.0% in 15 tests, and for maize and wheat which had 1% dyed grains, the recovery range was 0.3-1.5% in 30 tests. Thus accurate recoveries were obtained.

The PFS method of sampling is accurate because, unlike spear sampling, the whole bag is sample unit and the sample is obtained randomly, each grain having a chance of being picked.

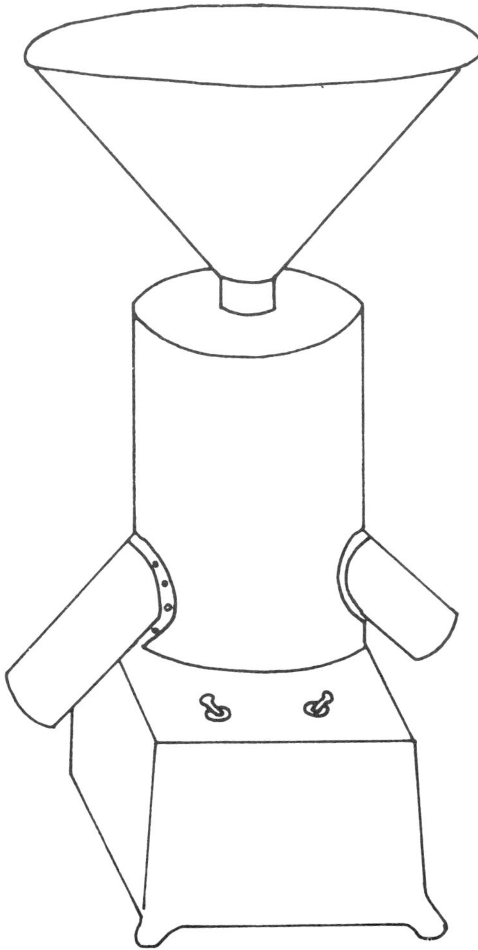

Fig. 23. Gamet divider.

Conclusions

Samples obtained from bagged produce must be both representative and random of that produce. Sampling using a spear is not random and does not result in a representative quantity of produce being taken. At farmer or trader level, coning and quartering do provide accurate results and samples of similar quality at marketing board or export level are best obtained with the PFS.

Fig. 24. Produce flow sampler.

Sieving, although not strictly a sampling procedure, can give accurate estimation of surface insect population. Subdivision of primary samples must be random and the Boerner, box, and Gamet dividers fill this function. However, the equipment for subdivision of samples is relatively sophisticated and is not always available. It may be more practical and almost as reliable to reduce samples by coning and quartering.

Literature Cited

1. JELIER, G. Sampling of grains, milled products, starch products, and potato starch. Int. Assoc. Cereal Chem. ICC Standard 101 (1970).

APPENDIX B

TABLES OF RANDOM NUMBERS AND THEIR USE

B. Drew and T. Granovsky

Sample selection by means of randomization is not an unorganized hit-or-miss process. It is a rather formal protocol-dictated process to assure that an intentional or unintentional bias will not be introduced during sample selection and sampling.

A random sample means that each and every unit (ears, plants, baskets, row, farm) in a population has an equal chance of being selected. It means that the selection of "good looking," or "typical," or "some of the good ones and some of the bad ones," or those within a convenient distance will be avoided. To select on such bases neglects the principle that each sample should have an equal chance of being selected. Any such selection, therefore, introduces bias. Random selection usually means that randomization must be done by the project planners and supervisors although it may be accomplished at the working level and situations may be classified by the state of knowledge into 1) where information about the size of population to be sampled is available before field-workers are sent out, or where field-workers are competent to randomize, and 2) where information about population sizes is not available in advance and field-workers are not competent to randomize.

In either situation, the only way to select at random is by a table of random numbers. Any other means simply will not give the total randomization that is provided by a table of random numbers.

A table of random numbers (see Table VII) should be used by a fixed procedure determined in advance. To do this, one should know in advance what units and how many are to be taken as the sample: ears of corn, bags of grain, farms lying on map coordinates, etc. The procedure for taking the sample also needs to be established in advance.

1. Plan the selection of elements to be sampled in advance. Decide what is to be selected: rows, bags in piles requiring predetermination along a three-dimensional grid, bags as they are moved for sampling, etc. Decide how many of these units are to be taken for the sample.

2. Number the units in any convenient way starting with 1 and going as high as necessary.

3. Use the table of random digits. Start at any point in the table and proceed to read off pairs of digits in any direction — up, down, sideways, diagonally.

4. Write down the pairs of digits as they occur. Skip any numbers that are repetitions, or that are bigger than the total number of units numbered in step 1.

5. When you have written down the number of units to be taken in the sample, stop.

6. Sample those units whose numbers have been listed.

7. Each time the table is used indicate the starting pair of digits by circling. Do not start at the same place again.

Cases Classified by Situation

1. Where units can be numbered in advance:
   Cribs on a farm
   Baskets in a building
   Houses in a village
   Stacks in a field
2. Where units are encountered sequentially:
   Bags being unloaded from a truck or boat

Table of random numbers 1 to 100 with no numbers repeated in each block of 25.

```
75 64 26 45 10    79 18 58 61 09    67 05 60 19 91    14 62 02 35 98    88 51 53 56 96
24 05 89 42 27    98 62 31 19 95    24 25 58 50 49    19 30 31 58 59    49 47 85 48 30
63 18 80 72 41    26 11 91 96 81    55 92 44 23 93    97 89 53 40 80    29 46 34 39 63
38 81 93 68 22    84 92 59 82 80    26 94 73 71 45    63 84 68 44 94    93 64 13 94 31
25 59 54 43 02    16 41 97 40 65    70 29 77 74 27    69 81 70 01 95    82 99 77 80 21

12 28 15 88 98    21 28 92 06 08    33 72 05 13 06    85 65 33 90 20    92 33 27 59 49
36 59 95 67 96    25 72 30 41 81    71 92 18 65 17    64 58 56 89 28    69 18 36 06 71
91 72 33 68 11    22 20 15 01 65    34 60 47 16 09    44 45 46 97 83    44 51 98 67 29
86 04 47 43 69    12 85 04 93 74    80 08 57 25 79    72 96 07 57 40    82 62 68 60 73
01 05 65 97 77    96 64 98 62 49    07 19 63 46 66    77 98 80 54 60    97 32 83 74 80

26 95 96 93 87    17 59 90 35 94    73 68 03 27 29    49 64 66 14 65    57 24 45 76 39
45 27 71 62 05    71 18 32 42 91    25 66 46 49 71    67 11 25 23 12    41 47 99 66 01
74 07 90 20 25    05 52 65 84 92    87 57 95 37 83    85 45 22 56 26    10 28 04 88 49
77 99 91 43 02    96 06 07 36 68    17 48 06 09 84    31 86 91 87 96    63 87 32 33 70
75 53 35 46 41    21 95 85 61 46    94 18 78 39 47    19 60 48 15 59    68 79 42 09 67

45 65 84 36 28    48 33 82 62 71    74 48 75 92 34    32 94 26 70 88    35 50 19 97 52
81 74 60 90 46    13 51 24 54 55    45 54 12 90 99    44 68 86 71 58    27 51 81 11 77
95 11 96 85 83    93 53 74 52 97    79 53 21 41 44    45 81 02 38 07    38 07 80 89 56
29 40 82 33 86    67 95 43 41 89    05 52 17 31 13    82 61 78 57 40    84 39 57 63 78
79 14 32 21 09    32 27 02 70 20    61 47 24 42 76    77 27 99 36 15    36 98 08 40 53

51 46 23 17 11    93 35 70 37 86    26 23 64 88 17    17 78 95 93 83    65 23 90 78 55
98 75 60 99 89    91 18 20 27 74    31 82 01 32 97    97 43 21 87 82    33 28 10 56 98
15 97 42 56 79    08 58 79 40 31    37 19 20 58 41    41 86 66 54 45    08 76 89 86 32
06 16 35 93 26    36 97 26 17 71    74 95 89 08 50    50 62 48 46 26    24 95 93 01 64
54 43 55 21 74    47 59 75 03 57    63 38 02 51 77    77 76 65 08 92    72 29 35 06 85

66 31 33 83 19    15 01 38 69 66    77 83 87 16 45    04 07 72 32 08    53 91 03 48 49
06 07 88 09 61    19 29 39 18 16    76 48 53 81 12    61 39 87 60 33    84 75 78 22 55
57 01 84 02 27    11 14 47 20 44    22 34 90 86 79    89 68 71 46 77    08 76 89 86 32
47 08 89 24 85    87 13 48 68 94    07 70 88 03 36    75 92 73 05 56    62 37 77 34 42
17 05 93 51 30    82 49 61 45 31    91 55 23 11 89    53 15 34 76 78    33 41 99 79 43

15 19 85 03 11    81 76 26 77 13    73 75 64 47 85    08 61 70 03 25    90 92 94 98 97
91 64 24 16 46    23 44 70 47 17    10 70 43 35 56    67 73 71 90 57    37 34 54 95 35
70 09 43 21 61    24 74 07 96 33    08 42 19 74 12    09 27 77 23 17    93 43 14 38 15
62 94 51 92 60    49 25 15 85 34    86 09 11 03 96    47 54 02 32 76    75 13 76 32 03
53 13 59 22 82    87 37 94 62 65    18 40 14 38 71    41 55 14 50 28    62 74 08 31 58

93 59 48 96 88    04 83 14 84 53    45 70 37 18 05    79 14 45 55 46    28 55 36 35 77
58 14 07 89 30    51 76 38 05 32    13 01 23 63 33    24 73 13 21 16    46 78 20 67 32
47 40 60 22 29    52 16 70 44 19    46 41 93 73 78    68 88 42 02 28    66 17 83 37 38
28 02 81 52 80    56 08 63 06 22    35 50 32 75 22    66 69 65 97 35    87 65 33 29 10
69 24 61 41 42    24 73 45 55 46    47 21 95 09 62    86 67 29 74 54    95 14 74 72 79
```

Farmers coming to market
Farms located along a road
3. Where units can be designated by coordinates:
Map coordinates
Three-dimensional (a pile in a warehouse)

Special Instructions for Map Coordinates

Map Coordinates Method 1 (Preferred Method)

Number every grid point on the map. Leave out grid points that are inaccessible. Choose pairs of random digits as given earlier. If there are more than 100 grid points, follow the same procedure but use triples of random digits.

Map Coordinates Method 2 (Alternative Method)

Consider the vertical (north-south) coordinates to be units to be sampled. Number them from 1 up and use random numbers to choose as many coordinates as are needed in the sample. In this case *do not* skip repetitions.

Then consider the east-west (horizontal) coordinates to be the units. Number them from 1 up and use random numbers to choose as many coordinates as are needed. As each coordinate is chosen, pair it with the next unused one of the N-S coordinates from the first set. Repetitions are only skipped if they are paired with the same N-S coordinate.

APPENDIX C

MOISTURE METERS

Part 1

Guidance in the Selection of Moisture Meters for Durable Agricultural Produce[11]

T. N. Okwelogu

The market for moisture meters is both specialized and growing, and there is a need for special attention to the selection of meters. The manufacturer aims to reach as many possible users as he can, while the prospective buyer wants to know about as many meters as he can before investing in any model. Over the years 1966-70 enquiries about moisture meters have been received at The Tropical Stored Products Centre at the rate of approximately 100 a year. These enquiries have varied from wanting to know if a particular meter had a supply address in the locality of the enquirer, to seeking advice on what meter should be used for a specified purpose.

This statement is not a treatise on moisture meters, but an attempt to help the prospective buyer dealing with durable agricultural produce to determine which moisture meter best meets his requirements.

Sources of Information

The three principal sources of information available to the prospective users are 1) newspapers, magazines, and journals, 2) manufacturers' brochures, and 3) organizations in a position to give unbiased information about moisture meters.

Some newspapers, magazines, and journals, which occasionally contain information about meters, include the Financial Times, Electronic Age, and Power Farming. While manufacturers are always helpful in supplying data about their own range of meters, information about a wider range of meters will be more likely obtained from organizations having unbiased interest in these instruments. Examples of such organizations are 1) Tropical Stored Products Centre (Tropical Products Institute), Slough, England, 2) Grain Storage Department, Pest Infestation Control Laboratory, Ministry of Agriculture, Fisheries and Food, Slough, England, 3) National Institute of Agricultural Engineering, Wrest Park, Silsoe Beds, England, and 4) Grains Division, Agricultural Marketing Service, U.S. Department of Agriculture, Agricultural Research Center, Beltsville, MD 20705. Articles on moisture meters sometimes appear in the publications of these and similar organizations.

Tables VIII and IX give details of some available moisture meters, particularly how they can be obtained and the commodities with which they may be used. These details are based on information provided by the manufacturers of the meters.

---

[11]Adapted from Trop. Stored Prod. Inf. 21: 19 (1971).

TABLE VIII

Details of Some Available Proprietary Moisture Meters[a]

Meters Under Principles of Action	Power Supply	Test Speed	Accuracy (Within % MC)	Price Rating	Manufacturer/Supplier
CHEMICAL (C)					
C.1 Speedy	None required	Over 5 min	0.5	Under £50	Thomas Ashworth & Co. Ltd. Sycamore Avenue Burnley, Lancs, England
DRYING (D)					
D.1 X17 Agat	Mains	Over 5 min	0.3	Under £50	A.B.G.L. Jacoby Box 23014Y Stockholm 23, Sweden
D.2 Cenco Moisture Balance	Mains	1-5 min	0.2	Under £50	Cenco Instrumenten Mij, n.v. Konijnenberg 40, Post Box 336 Breda, Holland
D.3 Dynatronic IR Moisture Analyzer Mark II	Mains	1-5 min	0.2	Over £100	Lab-Line Instruments International Lab-Line Plaza 15th & Bloomingdale Aves. Melrose Park, IL 60160 USA
D.4 ts Crop Tester	Mains	Over 5 min	1.0	Under £50	Tower Silos Ltd. 2 Brock Street Bath, Somerset, England
D.5 Vacuum Moisture Tester	Mains	1-5 min	0.1	Over £100	Townson & Mercer Ltd. Croydon CR9 4EG, England
ELECTRICAL CONDUCTIVITY (Ec)					
Ec.1 KPM Aqua Boy	Battery	Under 1 min	0.2	£50-£100	K.P. Mundinger GmbH D-7253 Renningen, W. Germany

(continued on next page)

TABLE VIII *(continued)*

Details of Some Available Proprietary Moisture Meters

Meters Under Principles of Action		Power Supply	Test Speed	Accuracy (Within % MC)	Price Rating	Manufacturer/Supplier
Ec.2	Universal Moisture Tester	Self-germinating	1-5 min	0.2	Over £100	Burrows Equipment Co. 1316 Sherman Avenue Evanston, IL 60204 USA
Ec.3	Safe Crop Moisture Tester	Battery, Mains	1-5 min	0.5	£50-£100	
Ec.4	Agil Moisture Meter	Battery	Under 1 min	1.0-2.0	Under £50	Agil Ltd., Nicholson House Nicholson's Walk Maidenhead, Berks, England
Ec.5	Hart Moisture Meter K101, K103	Battery, Mains	Under 1 min	0.2	Over £100	Hart Moisture Meters, Inc. 400 Bayview Ave. Amityville, NY 11701 USA
Ec.6	'Hydraprobe' Copra Moisture Meter	Battery	Under 1 min	2.0	Under £50	Coe's (Derby) Ltd. Thirsk Place, Ascot Drive Derby DE2 8JL, England
Ec.7	Marconi Moisture Meter TF933B	Battery, Mains	1-5 min	0.5	£50-£100	Marconi Instruments Ltd. Longacre St. Albans, Herts, England
Ec.8	Protimeter Grainmaster	Battery	1-5 min	0.5	£50-£100	Protimeter Ltd. Field House Lane Marlow, Bucks, England
Ec.9	ScotMec-Oxley	Self-germinating	Under 1 min	1.0	£50-£100	Scottish Mechanical Light Industries Ltd. 42-44 Waggon Road Ayr, Scotland
Ec.10	Siemens Moisture Meter	Battery, Mains	1-5 min	0.5	Over £100	Siemens (UK) Ltd. Grt West House, Grt West Rd Brentford, Middx, England

(continued on next page)

TABLE VIII *(continued)*

Details of Some Available Proprietary Moisture Meters

Meters Under Principles of Action		Power Supply	Test Speed	Accuracy (Within % MC)	Price Rating	Manufacturer/Supplier
DIELECTRIC CONSTANT (Ed)						
Ed.1	Cera Tester	Battery	Under 1 min	0.3	£50-£100	A/S N. Foss Electric 39 Roskildevej, 3400 Hillerød, Denmark
Ed.2	Kappa-Janes Moisture Meter	Battery, Mains	1-5 min	0.5	Over £100	Kappa Janes Electronics 27 Stewart Avenue Shepperton, Middx, England
Ed.3	Burrows Moisture Recorder	Mains	Over 5 min	0.3	Over £100	Burrows Equipment Co. 1316 Sherman Ave. Evanston, IL 60204 USA
Ed.4	Lippke Moisture Meter FK-R-6	Mains	Under 1 min	0.5	Over £100	Paul Lippke K.G. 545 PO Box 1760 Neuwied, Germany
Ed.5	Wile	Battery	1-5 min	1.0	Under £50	OY Fima Ltd. Helsinki 70, Finland
Ed.6	Super-Matic Foss	Mains	1-5 min	0.3	Over £100	A/S N. Foss Electric 39 Roskildevej, 3400 Hillerød, Denmark
Ed.7	Transhygrolair	Battery	. . .	1.0	Under £50	Les Applications Industrielles de la Radio 236 Chemin des Vitarelles Tournefeuille (31), France
Ed.8	Steinlite Meters	Battery, Mains	1-5 min	0.3	Over £100	Seedburo Equipment Co. 618 West Jackson Boulevard Chicago, IL 60606 USA
Ed.9	Dole 300 Moisture Tester	Battery, Mains	Under 1 min	. . .	£50-£100	Eaton Yale & Towne Inc. Dole Division, 191 E. North Ave. Carol Stream, IL 60187 USA

(continued on next page)

TABLE VIII *(continued)*

Details of Some Available Proprietary Moisture Meters

Meters Under Principles of Action		Power Supply	Test Speed	Accuracy (Within % MC)	Price Rating	Manufacturer/Supplier
Ed.10	Cae Moisture Meter Model 919	Battery	Under 1 min	0.3	£50-£100	Canadian Aviation Electronics Ltd. Winnipeg 4, Canada
Ed.11	G-c-Wyndham Moisture Meter	Battery	Under 1 min	0.5-1.0	Under £50	E.J. Chapman & Co. Ltd. Martley, Worcester, England
Ed.12	C.D.C. Automatic Moisture Meters	Mains	Under 1 min	0.3	Over £100	Compagne des Compteurs (GB) Ltd. Terminal House, Grosvenor Gdns.
	Hyb 24, Hyb 25 Hyb 42, Hyb 43	Battery	Under 1 min	0.5	Over £100	London SW1, England
INTER-GRANULAR RH (H)						
H.1	Dip-Shaft Humidity Indicator	None required	Over 5 min	1.0	Under £50	Abrax Inc. 179/15H Jamaica Ave. Jamaica, NY 11432 USA
H.2	Quicktest Models 1 and 2	None required	Over 5 min	1.0	Under £50	Opancol Ltd. 10/11 Gamage Building Holborn Circus London EC1, England

[a]All information in this table has come from manufacturers. The exclusion of an instrument from this table does not necessarily imply the author's disapproval of its use with agricultural produce.

TABLE IX

Commodities and Some Candidate Moisture Meters[a]

Meter Ref. No.[b]	Moisture Meter	Range of Produce Moisture Content[c] (%)	Sample Size[d] (g)	Beans	Cashewnut	Cereals and Products	Cocoa Beans and Products	Coconut (Copra)	Coconut Desiccated	Coffee Beans	Cottonseed	Cowpeas/Peas/Grams	Dried Fruit and Nuts	Fishmeal/Shreds	Groundnut	Horticultural Seed	Milk Powder	Palm Kernel	Sesame or Benniseed	Soya Bean	Sunflower Seed	Tea	Tobacco
C.1	Speedy	0-50	S-M			+										+							
D.1	X17 Agat Moisture Meter	M	M			+																	
D.2	Cenco Moisture Balance	M	S-M			+	+			+	+					+	+			+			+
D.3	Dynatronic IR Moisture Analyzer	M	S-M			+	+			+	+					+	+			+			
D.4	ts Crop Tester	M	L			+																	
D.5	Vacuum Moisture Tester	M	S			+																	
Ec.1	KPM Aqua Boy	2-30	L			+	+	+		+	+		+									+	+
Ec.2	Universal Moisture Tester	9-40	L	+		+					+				+	+		+					
Ec.3	Safe Crop Moisture Tester	11-30	. . .			+				+					+					+			
Ec.4	Agil Moisture Meter	14-30	L			+																	
Ec.5	Hart Moisture Meter K101, K103	Variable	M-L			+			+				+		+					+			+
Ec.6	'Hydraprobe' Copra Moisture Meter	4-14	M					+															
Ec.7	Marconi Moisture Meter TF933B	4-25	S	+		+				+		+	+		+								+
Ec.8	Protimeter Grainmaster	10-35	M-L			+																	
Ec.9	ScotMec-Oxley	10-25	L	+		+	+									+							
Ec.10	Siemens Moisture Tester	3-45	. . .			+																	+
Ed.1	Cera Tester	8-30	M			+				+				+	+	+		+	+	+	+		
Ed.2	Kappa-Janes Moisture Meter	0.5-60	L	+		+	+			+		+					+					+	+
Ed.3	Burrows Moisture Recorder	6-35	L	+		+	+			+	+				+					+	+		

(continued on next page)

TABLE IX *(continued)*

Commodities and Some Candidate Moisture Meters

Meter Ref. No.[b]	Moisture Meter	Range of Produce Moisture Content[c] (%)	Sample Size[d] (g)	Beans	Cashewnut	Cereals and Products	Cocoa Beans and Products	Coconut (Copra)	Coconut Desiccated	Coffee Beans	Cottonseed	Cowpeas/Peas/Grams	Dried Fruit and Nuts	Fishmeal/Shreds	Groundnut	Horticultural Seed	Milk Powder	Palm Kernel	Sesame or Benniseed	Soya Bean	Sunflower Seed	Tea	Tobacco
																							Commodities (+)
Ed.4	Lippke Moisture Meter FK-R-6	0-20	L			+								+										+
Ed.5	Wile	10-34	M			+															+		
Ed.6	Super-Matic Foss	6-40	L	+		+				+		+			+				+	+	+		
Ed.7	Transhygrolair	9-40	L			+																	
Ed.8	Steinlite Meters	4-36	L	+	+	+							+		+	+							
Ed.9	Dole 300 Moisture Tester	6-30	M	+		+	+			+					+					+	+		
Ed.10	Cae Moisture Meter Model 919	9-25	L			+				+										+			
Ed.11	G-c-Wyndham Moisture Meter	10-30	L			+																	
Ed.12	C.D.C. Moisture Meters	5-35	L			+	+			+		+			+						+		
H.1	Dip-Shaft Humidity Indicator	M	L			+																	
H.2	Quicktest Models 1 and 2	M	L			+																	
	Least available number of meters			8	1	29	7	2	1	12	4	5	4	2	10	7	3	2	2	9	6	2	7

[a] All information in this table has come from manufacturers. The exclusion of an instrument from this table does not necessarily imply the author's disapproval of its use with agricultural produce.
[b] As used in Table VIII.
[c] M = Max. range encountered in practice.
[d] S = Small (under 10 g); M = moderate (10-100 g); L = large (above 100 g).

With every piece of information, it is important to ask the question: Is this information sufficient for a decisive opinion to be formed about the meter? Where the answer is no, further enquiries should be made.

Factors to Consider in Making a Choice

It can be seen from Tables VIII and IX and in Parts 2 and 3 of Appendix C that there are several meters for any specific purpose. For satisfactory selection, the following factors should be carefully considered:
1. Meter types and their implications.
2. Characteristics of the commodity.
3. Requirements of the work for which a meter is sought.
4. Business considerations.

Principles and Implications of Meter Types

Most manufacturers indicate the principles upon which the action of their meters is based. An appreciation of the implications of such principles will be of considerable value in deciding which of several meters will be the most suitable. The meters commonly used with durable agricultural products fall

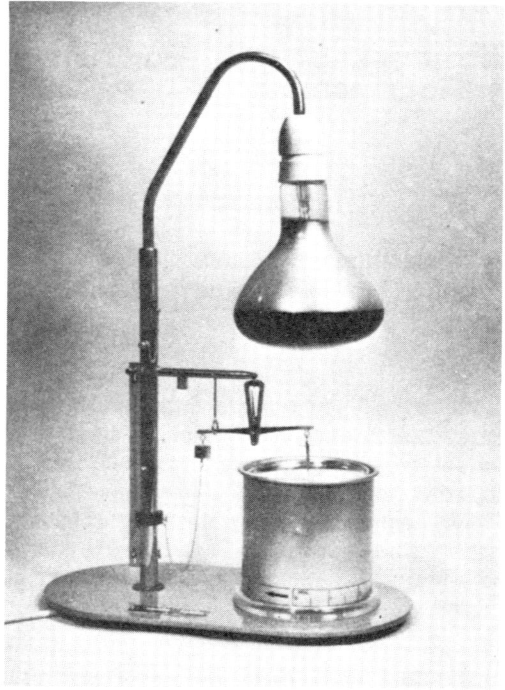

Fig. 25. Drying meter (Type D in Table VIII). A weighed, ground sample is heated by an infrared lamp for 10-20 min and the weight loss read directly in moisture content.

into five groups, according to the principles of their action:

1. Those involving chemical interaction between calcium carbide and the product water, with the evolution of acetylene gas, the pressure of which is subsequently measured.
2. Those involving heat-drying of the product, the attendant loss ascribed to evaporated produce water (Fig. 25).
3. Those involving measurement of electrical conductivity (or resistance) of the product, since the value of this property is related to the moisture content, within a suitable range of moisture contents (Fig. 26).
4. Those involving measurement of the dielectric constant of the product (or capacitance of the electrical system of which the product is a component), since the value of this property changes with the moisture content, within a suitable range of moisture contents (Fig. 27).
5. Those involving measurement of that atmospheric relative humidity which is in equilibrium with the product moisture, since, under equilibrium conditions, there is a definite relation between the moisture content of a product and the ambient relative humidity (Fig. 28).

Although it is tempting to try to list the advantages and disadvantages of the meter types, this approach is ineffective in providing buyers with adequate guidance. For example, although many resistance meters require a ground

Fig. 26. Electrical conductivity meter (Type Ec in Table VIII). Test cup is filled with whole or ground grain, which is then compressed and the meter read. The moisture content of wheat, rye, barley, and oats can be read directly off the scale, but tables are available for other commodities. Probe attachments are available for bagged grain.

sample, use a small sample, or test products with a relatively short range of moisture content, there are others in the same group which do not require the sample to be ground, which can test large samples (by using probes on whole sacks), or have an extended range of operating moisture contents. There are, nevertheless, certain outstanding group features to be noted: Heat-drying methods require a suitable source of power supply or fuel, which may not be available. Methods based on the evolution of acetylene gas require regular supplies of fresh calcium carbide, which is not a safe commodity to handle by post, because of the risk of explosion. Meters measuring the intergranular relative humidity require, first, a knowledge of the relation between the produce moisture content and the relative humidity of the intergranular air; secondly, a periodic check on their calibrations; and thirdly, in some cases, large quantities of produce which must have remained undisturbed for some time prior to testing.

The electrical meters are faster and, in the main, less demanding on calibration checks, but require skilled servicing. Also, they give less reliable readings outside the middle region of the range of moisture contents for which they are calibrated. The accuracy of the probe-type electrical meters is affected by variations in the pressure exerted by the produce on the electrodes, while the consistency of the readings of those meters which measure the dielectric constant is affected by inconsistent packing of the sample in the test chamber.

Attention has been focused above on the less favorable features of the meter groups mainly because they are more likely to be overlooked. Information on the merits of any meter will not normally be difficult to obtain, and Tables VIII and IX show the relative merits of the meters discussed in the present article.

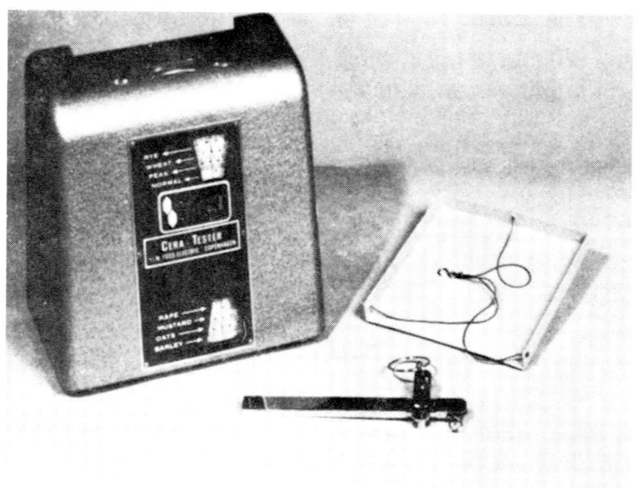

Fig. 27. Dielectric constant meter (Type Ed in Table VIII). A weighed, unground sample is poured into the cell. Moisture content is read on the scale or from table, and temperature correction from the thermometer on the back.

Characteristics of the Commodity

The commodity to be tested imposes a number of limitations, and these must be taken into account when considering the use of any meter. Perhaps the best way to do this is to answer questions such as the following:

First, is the chemical nature or any normal pre-treatment of the produce likely to interfere with the use of the meter? For instance, meters measuring

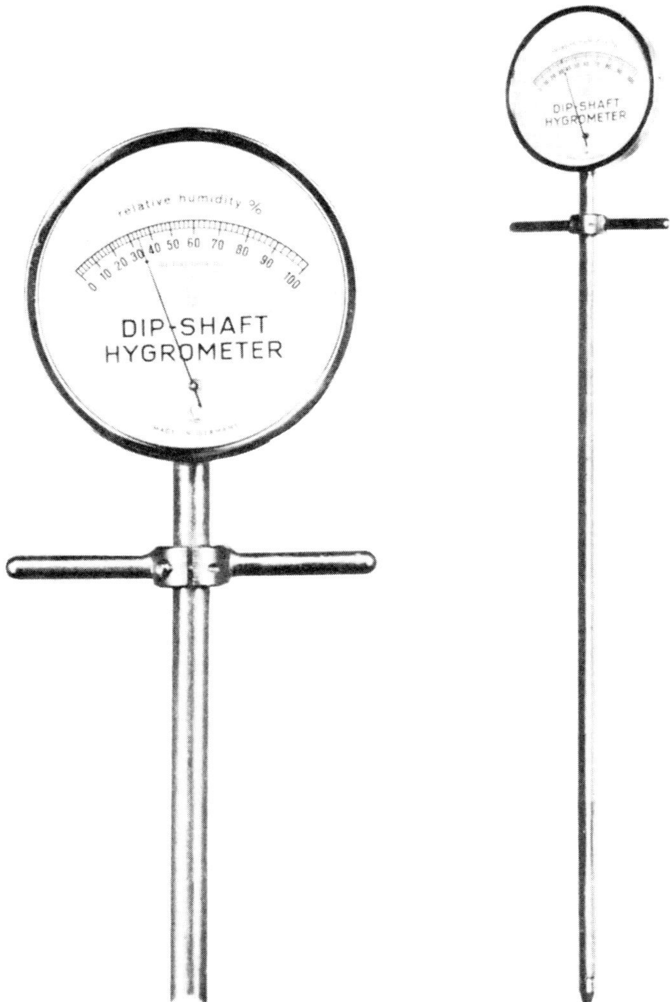

Fig. 28. Intergranular relative humidity meter (Type H in Table VIII). The indicator is pushed into the commodity in sack or bulk and allowed to equilibrate. Relative humidity reading is obtained and converted to moisture content using a calibration graph or table.

electrical conductivity may not be suitable for produce, like salt-fish, which will become highly conductive when damp. Again, for commodities like dried egg or milk, a heat-drying meter may not be suitable.

Second, is the moisture content to be measured outside the range for which the meter is calibrated? For example, very few electrical meters are known to be suitable for a product such as tea whose moisture content is normally required to be below 5%, that is, outside the range of moisture contents for which most electrical meters are calibrated.

Third, is the milling property of the produce incompatible with the effective use of the meter? For example, commodities such as macadamia nuts, palm kernels, copra, and cashew nuts are not amenable to grinding.

Fourth, are the unit size and shape of the produce likely to affect the efficient use of the meter? Construction of the meter may be such that it cannot be pushed into floury or powdery produce without hampering the measurement of moisture. Again, larger products like cocoa beans, unshelled groundnuts, cashew nuts, and pieces of illipe nuts (*Shorea* spp.) will present packing problems with some meters.

If the answer to each of the above questions is an unqualified no, then the meter may be considered suitable for the product. But a yes answer can make all the difference between a meter being chosen or rejected. In such a case, steps should be taken to see what, if anything, has been done to solve the problem, either by the manufacturer or by someone else.

Nature of the Situation Needing a Moisture Meter

In a summary of this kind, it is not easy, even if it is possible, to cover all the situations where the use of a moisture meter may be desired. However, such situations are likely to fall into one or the other of the following categories:

1. Knowing whether grain is at the right stage for harvesting.
2. The processing (eg, drying, milling, or storage) of foodstuffs.
3. Bulking or packaging for storage.
4. Commercial transaction, where moisture content is part of the basis for payments.
5. Produce inspection including loss estimates.

All the above situations require moisture meters which are not fragile, which are consistently accurate within limits acceptable for the particular purpose, and whose performance is little affected by the operating conditions of space, temperature, pressure, light, dust, or wind. They also require, to a certain extent, meters that are simple to operate, portable, and capable of taking remote measurements, as with probe-electrodes, or stem hygrometers, or that samples be taken of the material for laboratory testing.

Operational Considerations

The purpose for which the use of a meter is usually contemplated is twofold: to increase or improve productivity (that is, the flow of goods and services), and to ensure economical operations. The usefulness of the meter can be improved by employing one which can give results rapidly; for which spares and facilities for servicing or calibration are easily available; and which does not depend on sources of operating power that run out, break down, or become short in supply (eg, battery, mains supplies, gas, paraffin, and other fuel).

Economy of operation implies keeping to a minimum both capital and operating costs or increasing the return to unit cost. Additionally, even though it may have been purchased for a specific grain, the wider the range of commodities that a meter can test, the more flexible and economical may be its total use. Likewise, the less destructive a test is, the less will be the incidental loss of material caused by the use of a meter. Although this kind of loss may appear small, it must be realized that its magnitude will depend on how much produce is damaged at each test, and how many times such tests are performed on a given product.

Conclusions

Few meters, if any, can win the top position in every conceivable area of consideration, and there is no magic formula for choosing a meter. Where a choice has to be made, however, all known factors need to be considered. This implies having adequate information about as many meters as possible, and then carefully checking the meter descriptions against the requirements.

The buyer must have a knowledge of the commodity to be tested and the accuracy required of a determination of its moisture content; the availability of the meter, and the cost of operating it; the conditions under which the meter will be operated; the ease of obtaining spares and facilities for servicing or calibrating the meter; and the type of power supply required and available. When a provisional choice has been made, it is often advisable to obtain the meter on loan for trial before buying. This will make it possible to verify certain claims which may not be possible otherwise. For example, the buyer may discover that the meter does not give as "precise, error-free, and effort-less moisture measurement" as he was made to expect. He may discover, too, that although the meter is calibrated for rice, it in fact needs a different calibration for his own type of rice.

Choosing a moisture meter must be approached from both commercial and technical aspects, and requires a critical appraisal of many variables.

APPENDIX C

Part 2

Table of U.S. Department of Agriculture, Federal Grain Inspection Service List of Moisture Meters Used in the United States and Their Manufacturers, April 1978[a]

Name of Device	Principle of Operation	Manufacturer or Distributor
American Moisture Tester — Model M-20	Infrared heating — direct reading	American Farm Equipment Co. 340 E. Main St. Lake Zurich, IL 60047
Apollo Microwave Laboratory	Loss on drying using microwave energy	Apollo Microwave Products 6204 Official Road Crystal Lake, IL 60014
Auto-aquatrator	Karl Fischer Method	Precision Scientific Group 3737 West Cortland St. Chicago, IL 60647
Brabender, C.W. Rapid Moisture Tester	Thermobalance	C.W. Brabender Instruments, Inc. 50 East Wesley St. South Hackensack, NJ 07606
Brown-Duvel Moisture Tester	Distillation	Burrows Equipment Co. 1316 Sherman Ave. Evanston, IL 60204
		Gerber Industries P.O. Box 1387 Minneapolis, MN 55440
		Seedburo Equipment Co. 1022 West Jackson Blvd. Chicago, IL 60607
Burrows DMC-700	Dielectric	Burrows Equipment Co. 1316 Sherman Ave. Evanston, IL 60204
		Dickey-john, Inc. P.O. Box 10 Auburn, IL 62615
Burrows Moisture Recorder	Capacitance	Burrows Equipment Company 1316 Sherman Ave. Evanston, IL 60204
Burrows Safe Crop III Moisture Tester	Capacitance	Burrows Equipment Company 1316 Sherman Ave. Evanston, IL 60204
Burrows Model 400 (Radson) Moisture Meter	Capacitance	Burrows Equipment Company 1316 Sherman Ave. Evanston, IL 60204
Buhler MIAG Rapid Moisture Tester, Type MLI-400	Thermobalance	The Buhler Corporation P.O. Box 9497 1100 Xenium Lane Minneapolis, MN 55440
Cera-Tester	Capacitance	A/S N. Foss Electric Slangerupgade 69 DK 3400 Hillerød, Denmark

(continued on next page)

Name of Device	Principle of Operation	Manufacturer or Distributor
Delmhorst Moisture Detector	Conductance	Delmhorst Instrument Co. 607 Cedar St. Boonton, NJ 07005
Dickey-john DJ1S	Dielectric	Dickey-john, Inc. P.O. Box 10 Auburn, IL 62615
Dickey-john Forage Moisture Tester	Dielectric	Dickey-john, Inc. P.O. Box 10 Auburn, IL 62615
Dickey-john GAC-II	Dielectric	Dickey-john, Inc. P.O. Box 10 Auburn, IL 62615
Dickey-john GAC-III	Dielectric	Dickey-john, Inc. P.O. Box 10 Auburn, IL 62615
Digital Moisture Meter Model DM/6	R.F. Capacitive Measurement	Diversified Engineering, Inc. 2022 Sledd St. Richmond, VA 23220
Grain Quality Analyzer	Near IR	Neotec Instruments, Inc. 2431 Linden Lane Silver Spring, MD 20910
Higropant Moisture Meter	Conductance	National Instrument Co., Inc. 4119 Fordleigh Road Baltimore, MD 21215
Humidimetre Digital HD. 2000	Dielectric Fully Automatic	Cedem, Division Instrumentation Agricole Et Alimentaire 33-5 rue Jean Baptiste Charcot 92400 Courbevoie, France
Insto-I Moisture Tester	Dielectric	Dickey-john, Inc. P.O. Box 10 Auburn, IL 62615
Insto-II Moisture Tester	Dielectric	Dickey-john, Inc. P.O. Box 10 Auburn, IL 62615
KF-4B Aquameter System	Karl Fischer Method	Beckman Instruments, Inc. Scientific Instruments Div. P.O. Box C-19600 Campus Dr. at Jamboree Blvd. Irvine, CA 92713
KPM Aqua Boy MS-I	Conductance	Chatham International Corp. P.O. Box 377 Larchmont, NY 10538
Koster Crop Tester	Heating	Koster Crop Tester, Inc. 4716 Warrensville Ctr. Rd. North Randall, OH 44128
Marconi Moisture Meter Type TF-933C	Conductance	Marconi Instruments 100 Stonehurst Court Northvale, NJ 07647
Mettler LP 11	Infrared thermobalance	Mettler Instruments Corp. 20 Nassau St. Princeton, NJ 08540

(continued on next page)

Name of Device	Principle of Operation	Manufacturer or Distributor
Model G8R or Model G9	Radio frequency dielectric power loss factor	Moisture Register Company 6934 Tujunga Ave. No. Hollywood, CA 91605
Moisture Teller Model 276	Heating	Harry W. Dietert Company 9820 Roselawn Ave. Detroit, MI 48204
Motomco Moisture Meter — Models 919, 840, and 430	Capacitance	Motomco, Inc. 267 Vreeland Ave. P.O. Box 300 Patterson, NJ 07513
Ohaus Moisture Determination Balance	Infrared heating and balance	Ohaus Scale Corporation 1050 Commerce Ave. Union, NJ 07083
Optical Moisture Analyzer	Infrared absorption	Anacon, Inc. P.O. Box 416 Burlington, MA 01803
Pier Moisture Analyzer	Infrared Reflectance	Neotec Instruments, Inc. 2431 Linden Lane Silver Spring, MD 20910
Protimeter Grain Moisture Meters	Conductance	Cosa Corporation 17 Philips Parkway Montvale, NJ 07645
Quik-Test Moisture Tester	Dielectric	Dickey-john, Inc. P.O. Box 10 Auburn, IL 62615
Schenk Moisture Monitor	Capacitance and/or Conductance	Schenk Moisture Engineering R.R. 7, Box 78 Vincennes, IN 47591
Semi-Automatic Moisture Tester	Thermobalance	Haake, Inc. 244 Saddle River Road Saddle Brook, NJ 07662
Skuttle Moisture Meter	Conductance	Skuttle Manufacturing Co. Electronic Division Canfield, OH 44406
Steinlite Moisture Tester	Electronic Impedance	Seedburo Equipment Co. 1022 West Jackson Blvd. Chicago, IL 60607
Super-Conti	Capacitance Automatic Recording	A/S N. Foss Electric Slangerupgade 69, DK 3400 Hillerød, Denmark
Super-Matic I	Capacitance Print-out	A/S N. Foss Electric Slangerupgade 69, DK 3400 Hillerød, Denmark
T & M Vacuum Moisture Tester	Infrared vacuum thermobalance	Townson & Mercer, Ltd. Scientific Equipment Beddington Lane Croydon, England
Technicon InfraAlyzer	Near Infrared	Technicon Industrial Systems 511 Benedict Ave. Tarrytown, NY 10591

(continued on next page)

Name of Device	Principle of Operation	Manufacturer or Distributor
		Dickey-john, Inc. P.O. Box 10 Auburn, IL 62615
Universal Moisture Tester	Conductance	Burrows Equipment Company 1316 Sherman Avenue Evanston, IL 60201
902 Moisture Evolution Analyzer	Phosphorous Pentozide	E.I. DuPont Co. Instrument Products Div. Quillen Bldg. Concord Plaza Wilmington, DE 19898
No. 1210 Froment Moisture Tester	Mechanical Plunger — 9-volt Battery	N.J. Froment P.O. Box 758 Trenton, Ontario Canada

[a]This list was compiled from manufacturers responding to a USDA inquiry. Inclusion does not imply U.S. Government endorsement; omission does not imply disapproval.

APPENDIX C
Part 3

French Table of More Recent Moisture Meters with Acceptable Accuracy [a]

Brand [b]	Model	Address	Price (U.S. $) March 1978	Automatic Weighing	Digital Display	Printout	Calibration by		Sample Mass (g)	Power Source		Weight (kg)
							Manufacturer Only (Card)	Manufacturer or User (Keyboard)		Mains Supply	Batteries	
Automatic High Performance Apparatus Approved in France by the "Service des Instruments de Mesure"												
Cedem	HD 2000	33, rue Charcot 92400 Courbevoie (France)	3,250	Yes	Yes	Yes	Yes	No	400	Yes	No	43
Tripette and Renaud Dickey-john	Multigrain grain TR-Dj	39, rue Jean-J. Rousseau 75038 Paris Cedex 01 (France)	3,800	Yes	Yes	Yes	No	Yes	200-250	Yes	No	15
Foss-Electric	MK-II	69, Slangerup-gade DK 3400 Hillerød (Denmark)	3,300 approx.	Yes	Yes	Yes	Yes	No	170 approx.	Yes	No	15
Chopin Tripette and Renaud	ERAG-II	See above	1,470	No	No	No	No	No	10	Yes	No	13

(continued on next page)

Brand[b]	Model	Address	Price (U.S. $) March 1978	Automatic Weighing	Digital Display	Printout	Calibration by		Sample Mass (g)	Power Source		Weight (kg)
							Manufacturer Only (Card)	Manufacturer or User (Keyboard)		Mains Supply	Batteries	
					Other Apparatus							
Cedem	HD 1000	See above	480	No	No	No	Yes	No	400	No	Yes	4
Foss-Electric	M.K.-I	See above	1,990	No	No	Yes	Yes		250	Yes	No	20
	Cera-Tester		450	No	No	No	Yes		100	No	Yes	2
Chopin Tripette and Renaud	ERI-I	See above	670	No	No	No	. . .		10	Yes	No	10

[a]As summarized by J. L. Multon and G. Martin (see Chapter VI, Section F).
[b]All instruments use measurement of impedance of the grain, except the two ERAG ovens which use heat drying).

APPENDIX D

ASSESSMENT OF PROFITABILITY OF ALTERNATIVE FARM-LEVEL STORAGES[12]

M. Greeley

There have been relatively few attempts to assess the private profitability of alternative farm-level storage improvements. Yet without this evaluation there is no basis for choosing between alternative technologies.

The exercise below illustrates an approach to evaluating three important methods of storage improvement for Andhra Pradesh, India. In each case, we ascertain a benefit-cost ratio for each rupee invested by determining how many rupees are gained through grain saved by improving storage methods.

It must be emphasized that we are concerned here mainly with explaining the approach and that, for example, the levels of losses due to different causes given here are rough and are presented only as examples.

The three storage improvements, all designed by a local grain storage institute, are:

1. The domestic metal bin, manufactured by Andhra Pradesh State Agro-Industries Corporation;
2. the improved platform for the outdoor gade (bamboo basket); and
3. the improved base for the puri (large circular paddy-straw rope structure).

Improvements to the gade and puri are both designed to prevent access to rodents and groundwater migration. The puri is not fumigable but the gade can be fumigated successfully once a mud and dung coat is applied. The project has built over 30 gade improvements and 10 puri improvements. To make comparisons easy, all calculations are based on storage of one 75-kg bag of paddy. We are using loss-levels by cause[13] in the traditional stores of: rodents, 2%; insects, 2%; and molds, 1%, assuming that the maximum saving possible through storage improvement is 5%.

In addition, other values required are:

1. Initial construction costs both of the structure and the improved base/platform.
2. Annually recurring costs.
3. The price of paddy.
4. The effective life of the structures.

---

[12]This appendix is abstracted from a paper given in Coimbatore, Tamil Nadu, India, 1976, to a national meeting of engineers working on postharvest technology. The final report referred to is the IDS/IGSI Crop Storage Project report submitted to the Government of India in 1978.

---

[13]Comparison between the gade and the metal bin is unaffected by the relative importance of different causes because all three types of losses can be prevented in both. This is not true for the puri where fumigation is not possible. It may also be true that the importance of different causes of loss varies significantly between unimproved gades and unimproved puris as well as there being variation in the total percentage of losses, but the purpose here is to describe the method. The actual results are secondary, though it could be said that the improved gade-metal bin comparison is more realistic than comparing either one of these with the puri.

5. A discount factor.[14]

These values are given below:

Metal Bin. Currently priced at Rs 341 excluding transport and with a capacity of 10.5 bags, the cost per bag is Rs 32.5. Excepting fumigation there are no annually recurring costs and no platform costs. All three causes of loss are prevented.

The gade is a basket-type structure usually made from bamboo. Its cost depends on its capacity. Payments from the farmer to the basketmaker is in kind (not cash) at the rate of 2 kg of paddy for every 40 kg of capacity. To calculate the money value of a kind payment, we assume a price of Re 1 per kg of paddy. The cost of a 75-kg capacity structure, that is, one-bag capacity, is then equal to Rs 3.75. The cost of the improved platform is Rs 5.1 per bag. Total initial cost is therefore Rs 8.85.

The cost of the new mud coat each year is given as Rs 0.5 (based on an actual amount of Rs 8 for a 16-bag structure which is about average). The other annual recurring cost is fumigation. Total annual recurring cost is therefore Rs 1.25.

Improved Puri. The cost each year of the structure construction is approximately Rs 0.80 per bag after allowing for reuse of the straw. The structure is rebuilt each year. The cost of the improved base is Rs 4.2 per bag. Insect losses are not preventable because fumigation is not possible.

The life span of all permanent structure/platforms is conservatively estimated as 15 years.

The cost of fumigation (1 EDB ampule) is assumed to be Rs 0.75; one fumigation only is given at the time of initial storage.

Loading/unloading and cleaning costs are excluded since the puri is completely rebuilt each year and is loaded in the actual process of construction, but the estimated labor costs of loading (inseparable from construction) are roughly the same as for the other structures.

---

[14]A discount factor is a simple concept. It gives the relation between future cash flows and their present value. Asked to choose between a gift of Rs 100 now and Rs 100 in ten years' time, we would all choose Rs 100 now. To be willing to give up Rs 100 now, how much money would I require to be given in ten years' time? This depends on how much extra money I could earn in ten years with the Rs 100 invested, which in turn depends on the rate of return. This depends on the rate of interest. The discount factor works like a compound rate of interest. The value *now* of a Rs 100 in ten years' time is the amount of money I would have to invest now in order to have Rs 100 in ten years' time at a compound rate of interest. If I invest Rs 32 at a 12% rate of compound interest, its value in ten years' time is just under Rs 100; so the discounted present value of Rs 100 in ten years' time in this case is Rs 32. In valuing future costs or benefits to obtain their present value, we divide by a discount factor (the inverse of multiplying by a rate of interest). After one year an investment is worth $P(1+i)$, that is, the principal sum (P) plus the principal times the rate of interest. This sum which we call P_1 divided by $(1+i)$ equals P. Looking at the change after one year helps to understand the role of the discount factor. Rs 100 now at a 12% rate of interest equals $100 + Rs 12 after one year (100 + 100 × 0.12 = (P + P × i) = 112). We write this formula as $P(1+i)$. To find the original (present) value of that Rs 112 which we can call P_1, we simply reverse the process. Instead of multiplying by $(1+i)$ we divide by $(1+i)$. The present value is

$$\frac{112}{1 + 0.12} = 100, \quad \text{ie,} \quad P = \frac{P_1}{(1+i)}$$

Similarly, to reach the present value, after two years, we divide by $(1+i)^2$ and after three years by $(1+i)^3$. The value *now* of Rs 100 in 10 years' time is $\frac{100}{(1+i)^{10}} = Rs 32$

(where i = the proportionate rate of discount. In this case, the rate of discount is 12%, i = 12/100 = 0.12). We have assumed a discount rate of 12% simply because it is one used in some national planning exercises and it may reflect not too misleadingly the rate of return in alternative forms of investment.

TABLE X

Discounted Rupee Values of Benefits/Costs for 1 × 75 kg Bag Paddy (from Alternative Improvements)

Year	Discount Factor	Metal Bin				Improved Gade				Improved Puri			
		Money Costs	Discounted Costs	Money Benefits	Discounted Benefits	Money Costs	Discounted Costs	Money Benefits	Discounted Benefits	Money Costs	Discounted Costs	Money Benefits	Discounted Benefits
0	. . .	33.25	33.25	. . .	. . .	10.1	10.1	. . .	. . .	5	5	. . .	. . .
1	1.12	0.75	0.67	3.75	3.35	1.25	1.12	3.75	3.35	0.80	0.90	2.25	2.01
2	1.25	0.75	0.60	3.75	3.00	1.25	1.00	3.75	3.00	0.80	0.64	2.25	1.80
3	1.40	0.75	0.54	3.75	2.68	1.25	0.89	3.75	2.68	0.80	0.57	2.25	1.61
4	1.57	0.75	0.48	3.75	2.39	1.25	0.80	3.75	2.39	0.80	0.51	2.25	1.43
5	1.76	0.75	0.43	3.75	2.13	1.25	0.71	3.75	2.13	0.80	0.45	2.25	1.28
6	1.97	0.75	0.38	3.75	1.90	1.25	0.63	3.75	1.90	0.80	0.41	2.25	1.14
7	2.21	0.75	0.34	3.75	1.70	1.25	0.57	3.75	1.70	0.80	0.36	2.25	1.02
8	2.48	0.75	0.30	3.75	1.51	1.25	0.50	3.75	1.51	0.80	0.32	2.25	0.91
9	2.77	0.75	0.27	3.75	1.35	1.25	0.45	3.75	1.35	0.80	0.29	2.25	0.81
10	3.11	0.75	0.24	3.75	1.21	1.25	0.40	3.75	1.21	0.80	0.26	2.25	0.72
11	3.48	0.75	0.22	3.75	1.08	1.25	0.36	3.75	1.08	0.80	0.23	2.25	0.65
12	3.90	0.75	0.19	3.75	0.96	1.25	0.32	3.75	0.96	0.80	0.21	2.25	0.58
13	4.36	0.75	0.17	3.75	0.86	1.25	0.29	3.75	0.86	0.80	0.18	2.25	0.52
14	4.89	0.75	0.15	3.75	0.77	1.25	0.26	3.75	0.77	0.80	0.16	2.25	0.46
15	5.47	. . .	. . .	3.75	0.69	. . .	. . .	3.75	0.69	. . .	. . .	2.25	0.41
Total	. . .	43.75	38.23	56.25	25.58	27.60	18.40	56.25	25.58	16.20	10.49	33.75	15.35

Note: The costs for any one year occur 12 months before the benefit is realized, ie, the paddy is stored intact for one year. If it is assumed that the paddy is removed regularly over the storage season, then half the annual increase in the discount factor is subtracted before dividing the money benefits. A similar adjustment can be made if the total storage period is less than 12 months.

The price of paddy is assumed to be 1 Re per kg. The discount rate is assumed to be 12%. It is assumed also that no credit has been taken to purchase any of the structures so no loan or interest payments are due.

The costs are as above and the benefits over the 15 years' life of the structures are measured by the grain saved:

$$\text{Rodents } 2\% = \text{Rs } 1.5 \text{ undiscounted}$$
$$\text{Insects } \; 2\% = \text{Rs } 1.5 \text{ undiscounted}$$
$$\text{Molds } \; \; 1\% = \text{Rs } 0.75 \text{ undiscounted}$$

From the totals at the bottom of Table X the discounted benefits/cost are the money benefits/costs divided by the discount factor over a 15-year period.

The discounted benefit-cost ratios are as follows:

$$\text{Metal bin } 25.58{:}38.23 \qquad = 0.67{:}1$$
$$\text{Improved gade } 25.58{:}18.40 = 1.39{:}1$$
$$\text{Improved puri } 15.35{:}10.49 = 1.46{:}1$$

The importance of discounting is shown in the case of the metal bin. Without discounting the benefit-cost ratio is 1.29:1 (51.25:43.75), which implies that for every rupee invested, a return of Rs 1 29 paise can be expected, whereas after discounting we obtain a return of only 67 paise, for a loss of 33 paise. We must emphasize again that the loss-levels given are assumed only for convenience in illustrating the approach.

The same approach can be easily adapted to include additional factors such as risks of fire, flood, and theft or the use of different prices for (a) different uses of stored grain, or (b) different removal patterns. An important additional factor very relevant in some states now for the metal bin is the cost of credit. Further refinements can be introduced by examining how sensitive the results are to changes in the parameters (eg, different price levels). Indeed, this is an important exercise if the values used are at all uncertain. Some subjective factors such as the preference for a modern metal bin or contrarily the reluctance to switch from a traditionally used structure are more difficult to incorporate.

In this exercise we have ignored the question of *actual* storage requirements based on production and disposal patterns. If a farmer wishes to store 100 bags of paddy, then the theoretical choice would be between 1 puri, 4 gades (average size of our improved gades is 25 bags though individual gades up to 160 bags exist) or 10 metal bins. Space constraints and possible scale economics (which have been ignored by using average costs) then become relevant; both factors work in favor of larger unit capacity structures. However, it is also likely that, all things being equal, the *percentage* of losses is inversely related to size. In other words, the potential gross benefits from improvements to small structures are greater. The list of additional factors is by no means exhaustive; particular regions, particular crops, particular use patterns, etc., will require giving different emphasis to one or another factor but these can be incorporated as needed and still allow meaningful comparisons through the benefit-cost ratio.

Finally, we should note that a parallel approach can be used to estimate "social" benefit cost ratios from an extension program for storage improvement though this involves including (a) additional costs of the extension program and the associated administrative overheads, and (b) a set of prices that reflects real social values rather than using direct market prices.

SELECTED REFERENCES

ADAMS, J. M. A bibliography on post-harvest losses in cereals and pulses with particular reference to tropical and subtropical countries. Trop. Prod. Inst. G 110 (1977).

ADAMS, J. M., and HARMAN, G. W. The evaluation of losses in maize stored on a selection of small farms in Zambia with particular reference to the development of methodology. Trop. Prod. Inst. G 109 (1977).

ASIAN PRODUCTIVITY ORGANIZATION. Training manual: Post-harvest prevention of waste and loss of food grains. APO Project TRC/1X/73, Asian Productivity Organization, UNIPUB (1974).

van BRONSWIJK, J. E. M. H., and SINHA, R. N. Interrelations among physical, biological, and chemical variates in stored-grain ecosystems; a descriptive and multivariate study. Ann. Entomol. Soc. Am. 64(4): 789 (1971).

BROWN, R. Z. Biological factors in rodent control. U.S. Public Health Service, Communicable Disease Center Training Guide — Rodent Control Series (1960).

CHRISTENSEN, C. M. Storage of cereal grains and their products. Am. Assoc. Cereal Chem.: St. Paul, MN (1974).

CHRISTENSEN, C. M., and KAUFMANN, H. H. Grain storage: The role of fungi in quality loss. Univ. Minn. Press: Minneapolis, MN (1969).

COTTON, R. T. Insect pests of stored grain and grain products. Identification, habits and methods of control (out of print). Burgess Pub. Co., Minneapolis, MN (1963).

FOOD AND FEED GRAIN INSTITUTE, KANSAS STATE UNIVERSITY. Grain storage and marketing short course outlines (in English-Spanish-French). A. Fundamentals. B. Grain inspection and grading. C. Handling, conditioning and storage. D. Sanitation. E. Marketing, operations and management. Mimeo. Dep. Grain Sci. and Ind., KSU, Manhattan (1976).

GREIFFENSTEIN, A. C., and PFOST, H. B. Moisture absorption of bulk stored grain under tropical conditions. Res. Rep. No. 6. Food and Feed Grain Inst., Kansas State Univ., Manhattan (1974).

HALL, D. W. Handling and storage of food grains in tropical and subtropical areas. FAO Agric. Devel. Paper No. 90 (1970).

IDRC. Rice postharvest technology, ed. by E. V. Araullo, D. B. De Padua, and M. Graham. International Development Research Centre, Ottawa, Canada (1976).

LINDBLAD, C., and DRUBEN, L. Small farm grain storage. Action/Peace Corps, Program and Training Journal, Manual Series No. 2, Washington, DC, or Volunteers in Technical Assistance, Vita Publications, Manual Series No. 35E (1976).

MONRO, H. A. U. Manual of fumigation for insect control (2nd ed.). (In English-French-Spanish.) FAO Agric. Devel. Paper No. 79 (1969).

MUNRO, J. W. Pests of stored products. Hutchinson & Co., Ltd.: London (1966).

PEDERSEN, J. R., MILLS, R. B., PARTIDA, G. J., and WILBUR, D. A. 1974. Manual of grain and cereal product insects and their control. Dep. Grain Sci. and Ind., Kansas State Univ., Manhattan (1974).

PHILLIPS, R., and UNGER, S. G. Building viable food chains in the developing countries. Special Report No. 1. Food and Feed Grain Inst., Kansas State Univ., Manhattan (1973).

PINGALE, S. V., KRISHNAMURTHY, K., and RAMASIVAN, T. Rats. Food Grain Technologists' Research Association of India, Hapur (U.P.), India. Kapoor Art Press, Karol Bagh, New Delhi, India (1967).

RAMIREZ, G. M. Almacenamiento y conservacion de granos y semillas, 2a impresion. Compania Editorial Continental, S. A. Mexico, Espana, Argentina, Chile, Venezuela (1974).

SINHA, R. N. Uses of multivariate methods in the study of stored-grain ecosystems. Environ. Entomol. 6(2): 185 (1977).

SINHA, R. N., and MUIR, W. E. Grain storage: Part of a system. Avi Pub. Co.: Westport, CN (1973).

WORLD FOOD PROGRAM. Food storage manual. Part I, Storage theory. Part II, Food and commodities. Part III, Storage practice. Commodity and Technical Index. Prepared by Tropical Stored Products Centre, Ministry of Overseas Development, Slough, England (1970).

WYE, A. J. Selected bibliography on improving farm storage. Trop. Stored Prod. Inf. 21: 13 (1971).

INDEX